Understanding Modern Telecommunications and the Information Superhighway

For a complete listing of the *Artech House Telecommunications Library,*
turn to the back of this book.

Understanding Modern Telecommunications and the Information Superhighway

John G. Nellist
Elliott M. Gilbert

Artech House
Boston • London

Library of Congress Cataloging-in-Publication Data
Nellist, John G.
   Understanding modern telecommunications and the information superhighway /
John G. Nellist, Elliott M. Gilbert.
      p.   cm.
   Includes bibliographical references and index.
   ISBN 0-89006-322-2 (alk. paper)
   1. Computer networks.   2. Information superhighway.   I. Gilbert, Elliott M.
II. Title.
TK5105.5.N45   1999
384—dc21   99-10793
                                                                    CIP

British Library Cataloguing in Publication Data
Nellist, John G.
   Understanding modern telecommunications and the information superhighway. –
(Artech House telecommunications library)
   1. Telecommunication   2. Information superhighway
   I. Title   II. Gilbert, Elliott M.
   621.3'82

   ISBN 0-89006-322-2

Cover design by Lynda Fishbourne
Design and graphics by Matthias Chan

© 1999 ARTECH HOUSE, INC.
685 Canton Street
Norwood, MA 02062

International Standard Book Number: 0-89006-322-2
Library of Congress Catalog Card Number: 99-10793

Contents

Introduction

The Digital Convergence

As we move from the industrial age to the information age, the digital convergence of communications, video, computers, film, graphics, text, and sound on instantaneous, interactive delivery systems will have an effect on absolutely everything. There is a transformation going on that is global and unalterable. The change from analog to digital data, pumped to us at mega speed is going to change the world. We'll have new businesses, dead businesses, and old businesses, doing new business in new ways. We are witnessing the birth and death of small companies and multinational corporations as well as a changed world order. Everyone has trouble keeping pace with these changes.

What is the Information Superhighway? The World Wide Web? What is cyberspace? We are certainly living in interesting times, but nobody knows for sure where we are going. Whenever a new technology is introduced, it is quite common for the "expert" to be pessimistic. (A few famous quotations from the past are noted throughout the book). Either "it won't work" or "there's no market for it."

The techno-literate always overestimate the speed at which the public will adopt technology, while they underestimate the value of the human touch. Just because we can book a flight on our personal computers does not mean the demise of the travel agent, provided the agent offers service and sensible advice.

But nobody should underestimate the public's ability to adapt to change.

Consider a few of the changes that society has absorbed during the last 50 years. People born before 1946 entered a world without commercial television, credit cards, contact lenses, and dishwashers. They predated not only CD players, personal computers, and fax machines, but FM radio and electric typewriters as well. Back then a chip was a piece of wood, hardware meant hardware, and software was an HB pencil. Now, almost overnight, this digital convergence, the Internet, and the Information Superhighway have the capacity to revolutionize everything.

The Technological Revolution

The technological revolution is producing a serious income and employment gap between workers with technical and problem-solving skills, and those without. This is splitting society into two-tiers of "knows" and "know-nots." The economic changes taking place in our society are causing high anxiety as they up-end entire industries and companies, as well as individuals. But just as the mass-production, machine-driven industrial revolution progressed from extraordinary turbulence to postwar economic boom, today's technological revolution contains the seeds of tomorrow's prosperity.

It is said that this revolution will create a whole new wave of jobs, rising incomes, an array of new products, and widespread affluence. This technological change is causing painful upheaval, but it is an unavoidable transition. Just as soon as the revolution settles down, likely early in the next decade, it will ignite a sustained economic boom similar to the storied 1950 to 1970 era.

The U.S. Telecommunications Market

In the United States, the Clinton administration has shown great enthusiasm for the creation of an Information Superhighway across America. This national information infrastructure (NII), as it is called by the White House task force, will be an invisible, seamless, dynamic web of networks and information resources that will simultaneously carry limitless amounts of information in a variety of formats including voice, video, text, and multimedia to unlimited locations.

The Federal Communication Commission (FCC) and the National Telecommunications and Information Administration (NTIA) are trying to sort out the responsibilities for building the NII. The NTIA calls itself the primary U.S. federal agency working toward the definition and development of the NII, which is more commonly referred to as the Information Superhighway.

According to the NTIA, the NII will be a network that links people, businesses, schools, hospitals, communities, and governments, allowing them to communicate and exchange voice, video, and data information by using computers, telephones, radios, and other devices. The concept of the NII encompasses a wide range of telecommunications equipment, services and transmission media. The technology necessary for the NII includes, among other things, electronic cameras, computers, televisions, fiber optic transmission lines, microwave links, satellite systems, wireless networks, cellular telephones, pagers, and facsimile machines.

The NII will integrate and interconnect these physical components to provide a nationwide information conduit, accessible by everyone.

This issue has captured the interest of telephone companies and cable television companies throughout North America.

Tele-Communications Inc. (TCI), one of the largest American cable television companies, has begun to build a vast cross-country lightwave network capable of delivering 500 channels of entertainment and electronic information to its subscribers. In Orlando, Florida, Time Warner has created an interactive home entertainment and communication network for 4,000 subscribers who will be able to call up a wide range of movies on demand, interactive video games, and home shopping. Eventually, Time Warner wants to offer personal communications services (PCS), a mobile, wireless telephone system, with direct access to long distance telecommunications companies. Time Warner has recently merged with Turner Broadcasting. Turner's film library includes all MGM/UA, RKO, and pre-1950 Warner Bros. films.

The Regional Bell Operating Companies

The most important issue facing the Regional Bell Operating Companies (RBOCs), or "Baby Bells," created after the breakup of the Bell system in 1984, is which industry will dominate the provision of new multimedia services to the American home. As a result, US West has spent $11 billion to take over Continental Cablevision, the third largest U.S. cable system. In addition, it owns a 25% stake in Time Warner. Pacific Bell is spending $16 billion to create a multimedia communications superhighway across California. Nynex has invested $1.2 billion in Viacom, a cable television firm. Viacom recently acquired control of Paramount Communications Inc., owner of the last big independent film and television production company.

The fear of competition in the new liberalized markets is the principal cause of the rash of mergers and strategic alliances in the telecommunications industry.

A recent example is the merger between Bell Atlantic and Nynex. This created the second largest U.S. telecom company, capitalized at about $51 billion in 1997. Two other Baby Bells, SBC Communications and Pacific Telesis, have merged to create a $50 billion telecom giant.

In 1998, SBC Communications announced it would buy Ameritech Corp. for $62 billion, the largest merger in telecommunications history. The merger would make the new SBC the largest U.S. local telephone company.

The deals were triggered by the passage of the 1996 U.S. Telecommunications Act. This radical measure removes the regulatory barriers that have separated the telephone, cable, and broadcast industries. In particular, it allows long distance operators such as AT&T, MCI, and Sprint to compete in regional markets, and local operators like the Baby Bells to compete in the long-distance market.

Meanwhile, MCI, the second largest long distance company in the United States has been taken over by WorldCom Inc. The merged company will be the world's fourth largest telecommunications company behind Deutsche Telecom, AT&T, and Nippon T&T.

Also in 1998, AT&T Corp. agreed to buy TCI, the cable television giant, in a stock swap worth $48 billion. AT&T plans to deliver telephone service to two million homes over the cable television lines of TCI.

U.S. telecommunications companies continue to race for partners as the voice and data business becomes increasingly global and the Internet blurs country borders.

AT&T Corp. and British Telecommunications have announced that they will merge their international operations in a $10 billion global partnership. AT&T's new international partnership is the latest move to assemble a bold strategy to transform the telecom giant.

Bell Atlantic Corp. and GTE Corp. are holding merger talks to create a combined company with revenues of about $53 billion and they will control about one-third of the local telephone market. A combined GTE-Bell Atlantic would result in a company with local phone operations in 41 states. These deals, on the heels of other megamergers in the industry, put new pressure on the other regional carriers and long-distance companies to find mates before all the prime partners are gone.

Companies must be able to provide a wider range of services to customers who want the convenience of having their telephone, Internet, wireless, and paging service from one provider and on one bill. Traditional voice telephone carriers are increasingly becoming data companies because of the explosive popularity of the Internet and data services. Corporate clients also require a more sophisticated level of service as their own businesses become more global and more fast paced and competitive.

Established companies such as BellSouth Corp. and emerging companies such as Qwest Communications International Inc. and Level 3 Communications Inc. are seen as potential takeover targets.

The Canadian Telecommunications Market

In Canada, the Canadian-Radio-television and Telecommunications Commission (CRTC) has given telephone companies permission to apply for licenses to enter the cable television market. At the same time, the cable television companies can now compete in the local telephone market. A program designed to guide the route of the Information Superhighway in Canada was launched in 1994 by Canada's major telecommunications companies. The vision behind the program is a national, multimedia communications network linking all Canadians. If the railway was the national dream to link Canada in the 19th century, this program aspires to connect the country to the Information Superhighway now and into the 21st century. To make this dream a reality, the companies must make a number of changes to their own equipment and systems. This includes an $8 billion upgrade program over 10 years, as well as a $500 million enchancement program over 6 years to provide seamless national connectivity. The upgrade includes rebuilding the country's local telecommunications networks with fiber optic cables and asynchronous transfer mode (ATM) switches to permit the delivery of broadband multimedia services.

By the 2005 completion date, the telephone companies expect Canadians to be able to choose from a number of services. Some of the suggested possibilities would allow doctors in smaller locations to consult with specialists in major centers and share visual information about patients, using video conferencing. Also students and teachers would share resources electronically over thousands of miles; or consumers, using their telephone, televisions, or home computers, could take part in university courses or shop in electronic malls.

Two of the companies, BC Telecom in British Columbia and Telus Corp. in Alberta, have agreed on a $6.5 billion merger. The deal will create a company that could compete nationally with Bell Canada, the nation's largest phone company.

Determined not to be left behind by the telephone companies, Rogers Communication, Canada's largest cable television company and owner of one of the four major Canadian PCS networks, has formed a consortium with other cable firms to spend $6 billion over the next 5 years to upgrade and standardize technology, provide high-speed Internet access, and take on the major phone companies in the local calling market. The consortium will also set common standards and undertake group purchasing for large orders of new equipment, such as digital video set-top boxes.

Conclusion

A new 21st century marketplace is forcing the merger of telecommunications, television, computers, consumer electronics, publishing, and information services into a single interactive information industry. This will enable us to tap into and expand our vast resources of creativity and knowledge and will usher in a period of economic growth and prosperity fueled by this mega-industry, estimated to reach $3.5 trillion worldwide within the next decade.

1 From Telegraph to Terabits

Everything that can be invented has been invented.

— Charles Duell, Commissioner of the
U.S. Office of Patents, urging President
William McKinley to abolish his office, 1899.

In the Beginning

When we speak of telecommunications, we speak of communications over a distance, having added as a prefix the Greek root *tele*, meaning distance.

The first telecommunications media were mirrors, smoke signals, jungle drums, and semaphores, which represented each character to be transmitted by the position of mechanical arms. These were all extremely limited in both distance and maximum transmission speed.

The Wire Telegraph

The first telecommunications transmission by the electrical telegraph opened an era of invention and improvement of electrical telecommunications technology. This era boasts wired and wireless communication channels and switching centers that can be internetworked to provide a seamless global network.

The first electrical telecommunications were transmitted by the wire telegraph, invented by Samuel Morse in the 1830s. The telegraph utilized the interruption or reversal of a direct current electrical circuit to cause a remote sounder to produce a click at the beginning and end of each interruption or reversal. A short interval between clicks was called a "dot," and a longer interval (about three dot intervals in length) was called a "dash." Messages were transmitted by sending dots and dashes in conformance to the Morse code, which represented letters of the alphabet, numbers, and punctuation marks by combinations of dots and dashes. As many as six dots and dashes in combination were used to represent one character, and maximum sounder speed was around four clicks per second, resulting in a maximum transmission rate of about 20 words per minute. The messages were decoded by telegraph operators and initially recorded as handwritten copy.

The telegraph had a tremendous impact on the United States and the world. For the first time ever, reports of events from great distances were available almost instantaneously. The railroads granted right-of-way to telegraph companies and negotiated service contracts with them.

The railroads used the telegraph systems to control the movement of trains, and the telegraph companies provided telegraph service to the public. The newspapers used the telegraph to gather news of happenings around the country. During the Civil War military orders, information on troop movements, and reports on the outcomes of battles were transmitted by telegraph.

In 1850, Western Union was formed in Rochester, New York, to provide message service to the general public over its privately controlled but publicly accessible network. By 1851, more than 50 telegraph companies were in operation in different regions of the United States.

Western Union profited from the wartime business. Over a short period of time it absorbed the independent telegraph companies and became one of the largest corporations in the United States.

In 1850, the first undersea telegraph cable was laid between France and England, extending wire line telegraphic communication internationally. With manual operation, the transmission rate was still slow.

Guglielmo Marconi demonstrated the feasibility of international wireless telegraphy in 1901 by transmitting messages between Europe and North America by electromagnetic waves. His transmitter consisted of a spark gap attached to an antenna, but did not include any provision for limiting the spectrum of the transmitted radio energy to a particular frequency. Later, refinements such as circuit elements that restricted the transmitted energy to a narrow band around a selected frequency, frequency selective circuits for the receivers, and more efficient antennas were added by other inventors. The circuits, which will pass only the frequency to which they are resonant, consisted of a combination of a capacitor and an inductor and are known as "tuned circuits."

Printing Telegraphs and the Teletypewriter

David E. Hughes was granted a patent for a printing telegraph in 1855. A continuously rotating wheel had letters of the alphabet, numbers, and other characters on it. Keys at the sending end were operated to send electrical impulses that would stop the wheel and print when the desired character was over a moving strip of paper. The Hughes teleprinter increased the maximum transmission speed to between 40 and 45 words per minute.

In 1874 J. M. E. (Emile) Baudot invented a code that represents each alphabetical character, number, punctuation mark, or control command by a group of five equal-length units. Each unit could be in one of two possible states: voltage present on the line representing a 1, or no voltage on the line, representing a 0. Alternatively, a 1 can be represented by a positive or negative voltage and a 0 by the opposite polarity. Systems that consist of only two digits are called binary systems, and bit is an abbreviation of the term "binary digit."

Baudot developed prototypes of teleprinters (also called teletypewriters) during the 1870s, and by the early 1900s teletypewriters that used a version of the Baudot code began replacing manual telegraphy. Start and Stop bits were added to denote the beginning and end of each character. An example of the 5-bit code for the letter "J" is shown in Figure 1-1.

The teleprinters operated in the range of 100 to 600 baud, which is equivalent to 20 to 100 words per minute, a significant improvement over manual telegraphy.

Figure 1-1   Letter "J" in 5-Bit Code.

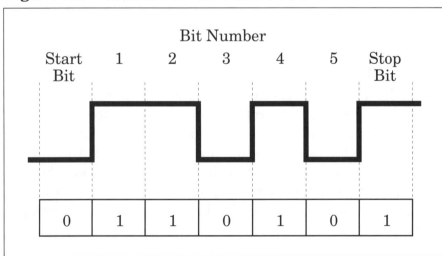

The Telephone

With the invention and demonstration of the telephone by Alexander Graham Bell in 1876, the electrical transmission of sound became possible. Sound waves are converted to electrical waves by a microphone at the sending end, as shown in Figure 1-2. The electrical waves are transmitted over a pair of copper wires and reconverted to sound waves at the receiving end. The electric waves are analogous to the sound waves. Thus, this type of transmission is called analog transmission.

The human ear can recognize sounds of frequencies between 30 and 16,000 Hz, with maximum sensitivity around 2,000 Hz. The maximum hearing range will vary from person to person and response to the higher frequencies diminishes with increased age. The human voice includes frequencies in the 200- Hz to 5,000-Hz range. Telephone company personnel determined that intelligible speech could be transmitted over circuits that pass the band of frequencies from 300 to 3,000 Hz

Figure 1-2   Analog Transmission.

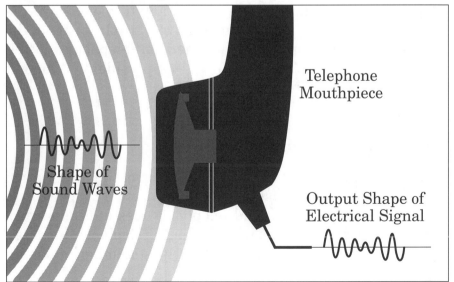

Telephone Mouthpiece

Shape of Sound Waves

Output Shape of Electrical Signal

5

and established this range as the requirement for voice circuits.

The Bell Telephone Company was formed in 1877 to produce telephones commercially. Western Union created the American Speaking Telephone Company, a competitor to Bell, in the same year.

In 1878, the Bell Company hired Theodore N. Vail, a former superintendent in the U.S. Post Office, to manage the organization, and it filed suit against the Western Union Telegraph Company for infringement of patents. In 1879, an out-of-court settlement was reached wherein Western Union agreed to give up the telephone business and Bell agreed to stay out of the telegraph business. In 1880, the American Bell Telephone Company was organized, and 2 years later Bell purchased the Western Electric Company to ensure a ready supply of telephone equipment.

The first long distance telephone service, between Boston, Massachusetts, and Providence, Rhode Island, began on January 1, 1881. Vail, now the president of American Bell, realized that the long distance lines were the key to monopolize the telephone network and began to acquire control. In 1885, the company was incorporated as the American Telephone and Telegraph Company (AT&T) with Vail as president. Vail left AT&T in 1887.

Telephone Switching

For a telephone network to be a useful part of the Information Superhighway, users must be able to establish a connection to any of the various nodes and telephone lines of the network quickly, accurately, and at will. Initially, telephone lines connected an individual telephone directly to another. Soon, in order to allow a telephone to be connected to one of many others, switching centers were established, where lines from any two telephones connected to the center could be interconnected. The switching was done manually by an operator by inserting a plug into a jack.

The switching centers became known as central offices, a term still in use today. Some of these central offices became so large that operators wore roller skates in order to move quickly between jacks.

Automatic Telephone Switching

In 1889, Almon B. Strowger, an undertaker in Kansas City, Missouri, suspected that, when potential clients asked the operator for an undertaker, they were connected to the telephone of a competitor down the street. He became more certain of this when he discovered that the operator was the wife of the owner of the other funeral parlor. As a result, Strowger worked on, invented, and patented in 1891 a mechanical switching device that could complete a connection under the control of the calling party. This switch, a step-by-step switch, completes a call by performing a sequence of operations in response to dial pulses. It is known as the Strowger switch after the name of the inventor or as the step-by-step switch.

The Strowger switch was first sold to independent telephone companies and was not installed by the Bell System until 1917. It was slow and noisy and required frequent maintenance. In 1921 the Bell system developed a panel switching system that was also slow and required frequent maintenance because it was a mechanical system. The Strowger switch and the panel switch had a maximum capacity of 10,000 lines. Any line could be accessed by dialing four digits. Additional capacity could be provided by adding another switch in tandem.

Voice Telecommunication Without Wires

In 1902 Reginald A. Fessenden, a Canadian-American, invented amplitude modulation (AM) of radio signals and transmitted the human voice over radio waves. Commercial radio telephony did not develop on a large scale until the availability of vacuum tube amplifiers. The first commercial radiotelephone stations were used to provide a link between Catalina Island, off the California coast, and the mainland. Radio broadcasting was demonstrated by Fessenden and E.F.W. Alexanderson on Christmas Eve 1906, when a program of speeches, song, and violin music was broadcast from a radio transmitter set up near New York City. The first commercial radio broadcasts began in 1920 by the Westinghouse station, now known as KDKA, in Pittsburgh, and by station 8MK, now WWJ, in Detroit. In 1926 commercial transatlantic radiotelephone service was initiated with a call made between New York and London.

In 1906 Lee DeForest patented the audion vacuum tube, which could amplify electrical signals 1,000 times or more. The tube improved the sensitivity of radio receivers, and telephone engineers discovered that it could amplify voice signals on telephone lines also, making possible transcontinental and trans- oceanic telephone calls. In 1915 coast-to-coast telephone service began in the United States.

AT&T and Western Union Separate

By 1907 J.P. Morgan had gained control of AT&T and persuaded Vail to return. In 1909 AT&T decided to go back into the telegraph business and purchased stock in Western Union. By 1910 AT&T had gained control of Western Union and Vail became president of that company. The independent telephone companies began to surpass AT&T in number of subscribers

8

after the AT&T patents expired. Vail and Morgan set out to make AT&T a monopoly. Morgan cut off credit to independent telephone companies, then Vail bought them. In 1912 the independent telephone companies protested to the U.S. Department of Justice that AT&T was operating in violation of the antitrust laws by refusing to interconnect with the independent companies, rate cutting to drive the independents out of business, and buying out competitors. Threatened with an antitrust suit, AT&T agreed in 1913 to dispose of Western Union and to allow independent telephone companies to interconnect with its network. AT&T became the voice carrier and Western Union became the "record" carrier of messages and documents.

Wire Line Carrier

In 1918 the long distance telephone lines consisted of open wire pairs strung on poles. One communication channel was carried by each pair. Frequency division multiplexing (FDM) was introduced that year in the form of carrier systems that allowed more than one voice communication channel to be transmitted over each pair. These systems utilized a group of carriers, each assigned a discrete portion of the wide-band frequency spectrum that could be transmitted over the line. Each carrier was modulated with one voice channel. At the receiving end, the voice channels were demultiplexed with frequency separation filters and demodulated. The carrier frequencies used were typically in the 36 to 500 kHz band, each occupied by 12 bidirectional channels, with 8 kHz of bandwidth assigned to each carrier in each direction.

Telecommunications Regulation

In 1921 the U. S. Congress passed the Graham Act, which exempted telephony from the antitrust act and allowed AT&T to acquire competing companies and legally create a virtual monopoly.

Congress created the Federal Communications Commission (FCC) and defined its powers in the Communication Act of 1934. The FCC was given jurisdiction over interstate and foreign commerce in communications, but not telecommunications within a state. Use of the radio frequency spectrum and licensing of radio transmitters were given to the FCC. Jurisdiction over telecommunications within each state was the province of the state's public utility commission.

Semiconductors Arrive

Transistors, made from semiconducting materials, were commercially introduced in 1948. They reduced the power and space requirements and increased the switching speed of electronic circuits by many orders of magnitude. Integrated circuits, which followed, combined millions of transistors and other circuit elements on a semiconductor chip. Silicon chips that contained complete computer processors were produced. The initial and maintenance costs of integrated circuit assemblies were much lower, making expansion of communications circuit capacity more attractive.

Advances in Switching

In 1938 the crossbar switching system was introduced. The crossbar switch used a matrix of electromechanical relays with input lines in vertical columns and output lines in horizontal rows. Connections were made by energizing the relay for a particular input line and a particular output line. The crossbar switches were less noisy and easier to maintain than the step-by-step switches.

Today, in North America and in many countries around the world, step-by-step and crossbar switches are being replaced by computer controlled, electronic digital switches with semiconductor integrated circuit switching elements. These switches can be programmed to install or delete a large list of user features such as call waiting, call forwarding, automatic identification of calling number, conferencing, and last number redial. Connections can be established in millionths of a second in response to multiple tones selected by push buttons on telephones or by automatic multitone dialers instead of the dial pulses produced by rotary dial telephones. Switches with capacities up to 100,000 lines are available.

Coaxial Cables

In order to further increase the capacity of transmission facilities between cities, telephone companies began installing coaxial cables that can carry many more channels than wire pairs. Coaxial cables can transmit signals with frequencies in the multimegahertz range with acceptable attenuation. A large number of telephone conversations can be transmitted over a coaxial cable by using FDM to divide the wide band of frequencies carried by the cable into many individual 4-kHz-wide telephone channels. A coaxial cable tube is shown in Figure 1-3.

A single tube commonly carries several thousand telephone channels. High-capacity coaxial cables have as many of 20 tubes which can carry a total of 100,000 telephone channels. The coaxial cable systems can also be used to carry television chan-

Figure 1-3   Coaxial Cable Tube.

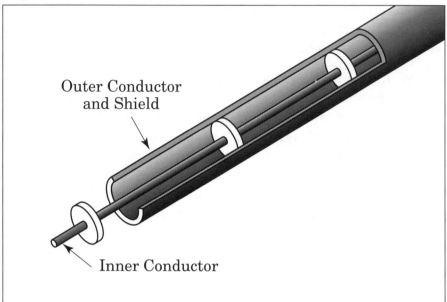

Outer Conductor
and Shield

Inner Conductor

nels. Each television channel requires the same amount of bandwidth as 1,200 voice channels.

Microwave Radio Arrives

The primary competition with coaxial cable for high capacity communications transmission in the 1950s through the 1970s was microwave radio. The first microwave relay station for long distance telephone communication was installed by the Bell system between Boston and New York in 1947. Microwave radio is so called because its operation with an extremely high-frequency carrier results in very short wavelength. One wavelength at a frequency of 6 GHz is 5 mm. The microwave carrier was modulated with analog voice channels separated by FDM and is known as "analog microwave radio." Microwave radio systems do not require continuous right-of-way for installation as do cables. Also, microwave radio towers are usually spaced 20 to 30 miles apart, so that a coast-to-coast system can be implemented with about 100 repeaters. A co-axial cable system of the same length requires about 1,000 amplifiers. One of the problems with microwave radio is interference due to the large number of microwave transmitters in each area and limited radio frequency spectrum.

Transistors, which were also introduced in 1947, along with integrated circuits that followed, reduced the power and space requirements of electronic circuits by several orders of magnitude. The initial and maintenance costs of integrated circuit assemblies was also much lower, making expansion of communications circuit capacity more attractive.

First Transatlantic Submarine Telephone Cable

The first transatlantic telephone cable (called TAT-1) was completed between Nova Scotia and Scotland in 1956. It utilized frequency division multiplexing of analog channels onto high-frequency carriers transmitted over copper pairs and vacuum tube amplifiers that were predicted to have a 20-year life. Subsequent generations of copper wire submarine cables used transistor amplifiers, and transmission capacity was increased to 4,000 voice frequency channels.

The Digital Transmission Era Begins

For the first half of the 20th century, electronic telecommunications technology was based predominantly on analog transmission and frequency division multiplexing. In 1962 Bell Labs introduced a digital transmission system that converted analog signals to digital form and used time division multiplexing

(TDM) to combine a number of digitized signals together for transmission over the same facility. A technique known as pulse code modulation (PCM), which was patented in 1939 by Sir Alec Reeves, is used in this system to convert the analog signals to digital form. The introduction of PCM revolutionized telephony transmission, and for the past 35 years millions of digital circuits and thousands of digital switches have been installed. While it was originally a technology for transmission over twisted wire pairs in cables, digital coding, PCM frame formats, and the digital hierarchy that have been developed are used for telecommunications transmission over copper cable pairs, digital microwave radio, and optical fiber.

PCM depends on three operations to produce the digital signal— sampling, quantization, and coding. A continuously varying analog signal, such as the electrical equivalent of a voice signal, is sampled by measuring the instantaneous amplitude at specific points in time. The sample is quantized by assigning it to the nearest voltage value of a set of discrete voltages reference scale.

Each discrete amplitude value is then coded into binary form, with the decimal value of the voltage represented by a digital word consisting of a combination of 7 binary digits (1s and 0s defined by the presence or absence of a pulse). An additional signaling bit is added to the word associated with a channel so that each sample is represented by an 8-bit word.

Sampling of a channel must be done at a rate at least twice the highest frequency of the channel to allow the channel to be reconstructed. Thus, an analog telephone channel with a bandwidth from 300 to 4,000 Hz must be sampled at least 8,000 times a second. Since 8 bits are required to represent the samples from a channel and each is sampled 8,000 times per second, it takes 64,000 bits per second to represent a channel.

Channels can be multiplexed together by assigning a unique time slot in a pulse stream to digital words from each channel. In standard North American digital transmission systems, 24 voice frequency channels are time division multiplexed into a 1.544-Mbps bit stream. This signal is called DS-1 (digital signal-level 1). The 8-bit words are placed sequentially in time slots and a synchronization bit is inserted after the word for the 24th channel. The 24 words plus the framing bit are known as a frame (see Figure 1-4), and this multiplexing process is called word interleaving. The 24 channels, each requiring 64,000 bits per second, produce a line rate of 1.536 Mbps. The framing bits add 8,000 bits per second for a total line rate of 1.544 Mbps. In the standard European transmission systems, 30 voice channels plus 2 channels carrying signaling and synchronization information are multiplexed to form a 2.048 Mbps bit stream. At the receiving end of a TDM/PCM transmission system, the channels must be demultiplexed and decoded to deliver a reproduction of the original analog signal.

Figure 1-4   PCM Frame.

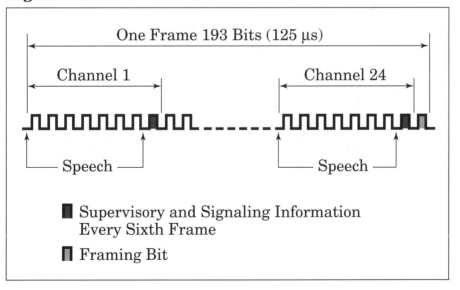

Higher level multiplexes are used to combine first order digital signals into higher order signals. Initially, there were three hierarchies of digital signal levels used in different parts of the world. In North America, the second order level, DS-2, was 6.312 Mbps, and DS-3 was 44.736 Mbps, as shown in Figure 1-5.

Recently, U.S. standards organizations, through the American National Standards Institute (ANSI) and the International Telecommunications Union (ITU) have agreed on compatible digital hierarchies, with gateway data rates, so that transmission systems from different countries can meet without requiring translation between systems. The data rates for the Synchronous Optical Network (SONET) and the international Synchronous Digital Hierarchy (SDH) are listed in Table 1-1.

One outstanding advantage of digital transmission is that computer and data signals are already digital and can be transmitted without conversion.

Figure 1-5   Digital Signal Hierarchy.

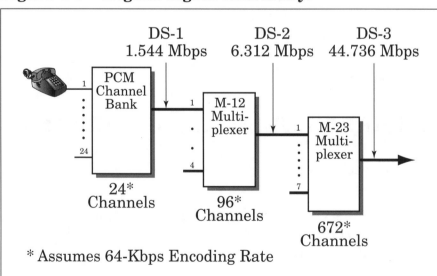

Table 1-1   SONET and SDH Line Transmission Rates.

ITU-T SDH*	SONET**	Optical Carrier	Data Rate	Voice Equivalent Channels
	STS-1	OC-1	51.84 Mbps	672
STM-1	STS-3	OC-3	155.52 Mbps	1,344 (Gateway)
STM-4	STS-12	OC-12	622.08 Mbps	5,376
STM-16	STS-48	OC-48	2.488 Gbps	21,504
STM-64	STS-192	OC-192	9.953 Gbps	86,016

*   SDH – Synchronous Digital Hierarchy
**   SONET – Synchronous Optical Network

Satellites Relay Communications

Satellites are microwave radio repeater stations within Earth's orbit. A satellite receives microwave signals from ground stations on the Earth's surface and retransmits them at a different frequency (to avoid interfering with the signals being received) to other ground stations thousands of miles away.

The first active telecommunications satellite, Telstar I, was launched on July 10, 1962. It was designed and built, and its launching paid for by the Bell telephone system. It carried telephone conversations, color television, and telephoto pictures between the U.S. and Europe. It was a low orbit satellite with an orbital velocity different from the rotational speed of the Earth's surface. Thus, its position with respect to places on the Earth's surface was constantly changing. The time that it was in view of an Earth station was less than one half hour. This precluded communications except when it was in a position between the ends of the desired communications path and not shadowed by the Earth.

The U.S. government passed the Communications Satellites Act of 1962, which allowed the Communication Satellite Corporation to pursue the commercial development of communication satellites. On April 6, 1965, Intelsat 1, or "Early Bird," was placed in an orbit 22,300 miles above the equator. At this altitude, the velocity required to maintain the satellite in orbit resulted in one complete orbit every 24 hours, allowing it to remain in a fixed position with respect to a point on Earth, with a view of an entire Earth hemisphere. This type of satellite is known as a geosynchronous satellite.

Communication satellites launched recently can relay data at rates of 50 Mbps and carry 60,000 voice frequency circuits or a mix of voice channels and television channels, with one television channel occupying the bandwidth required for 1,200 voice channels.

Community Antenna Television (CATV)

In many communities in the world, reception of broadcast television signals is not possible because of distance from the antennas or because the signals are blocked by mountains or buildings. In the 1960s entrepreneurs began to erect receiving antennas on mountain tops and other sites where the broadcast signals could be received and piped over coaxial cable to distribution centers, called headends, and then supplied over coaxial cable networks to individual subscribers who paid a fee for the service. This service became known as community antenna TV (CATV). The CATV companies were granted franchises in the communities where they operated. Later, television channels for CATV distribution were transmitted via satellite to large dish antennas at the headend site, and subscribers could receive more than 100 TV channels on the cable, paying additional fees for higher tiers of service.

Communications Through Glass Fibers

John Tyndall demonstrated in 1870 the principle of guiding light by means of internal reflections in the transmitting medium by showing that light could be bent around a corner as it traveled in a stream of pouring water. It took a series of developments nearly a century later to produce the components of a practical, working fiber optic communications transmission system. In 1966 Charles Kao and Charles Hockham of ITT England published a paper that stated that, if impurities in the glass could be reduced sufficiently, optical fibers could be produced with attenuation as low as 20 dB per kilometer (dB/km), and these fibers could be used as a transmission medium for light. By 1970 Corning Glass Works had produced fiber with losses under 20 dB/km, and by 1972 losses in laboratory samples had been reduced to 4 dB/km. Light sources and detectors were developed and prototype fiber optic transmission systems were constructed in the mid 1970s. In 1977 GTE installed field trial systems that carried commercial traffic between central offices in several cities in the United States and Canada, and AT&T installed a commercial lightwave system in Chicago.

The fibers available for these early systems had limited bandwidth because of the multimode dispersion, and the maximum span length between repeaters in the systems was limited because the bandwidth was inversely proportional to the fiber length. These systems transmitted TDM digital signals at rates compatible with the North American digital hierarchy. The early GTE systems operated at the DS-3 data rate of 44.736

Mbps, which was the digital equivalent of 672 voice frequency channels. At this data rate, unrepeatered link lengths were restricted to 10 km or less. The attenuation of the fiber at the operating wavelength was around 4dB/km.

The Breakup of Ma Bell

In 1982, 85% of all local telephone service and 97% of all long distance service in the United States was provided by AT&T. In 1982, Judge Harold Greene of the Justice Department issued his "Modified Final Judgment" that required AT&T to divest itself of 22 operating companies which were then allowed to provide only local telephone service. The Bell operating companies (BOCs) were organized into seven regional holding companies (see Figure 1-6). The act allowed specialized common carriers (SCCs) such as Sprint and Microwave Communications Incorporated (MCI) to build microwave networks and provide long distance service in competition with AT&T.

Advances in Fiber Optic Technology

Advances in fiber optic transmission technology in the 1980s produced light sources and detectors that could operate at longer wavelengths where attenuation was lower. Singlemode fiber, optimized for operation at longer wavelengths, was developed. Singlemode fiber did not exhibit the modal dispersion that limited the bandwidth of multimode fibers (see Figure 1-7). Singlemode connectors and splices with losses less than 0.1 dB were produced. Fiber optic spans with data rates of 565 Mbps on each fiber and 65 km long spans between repeaters were made possible by the new technology.

Carrier companies were formed that negotiated right-of-way

Figure 1-6   The Seven Regional Bell Operating Companies.

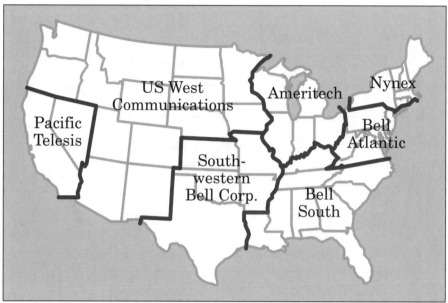

access with railroads, power utilities, and pipelines and installed fiber optic networks. Capacity of the networks was leased to the SCCs. MCI and Sprint replaced their microwave networks with optical fiber channels that were quieter and were not susceptible to weather and other effects that could interrupt service. AT&T added more fiber optic extensions to its network. Digital switches that provided interfaces to the network at standard rates in the digital hierarchy were introduced.

In 1988, the first fiber optic transatlantic telephone cable (TAT-8) was laid. This cable provided 40,000 telephone channels between the United States and the United Kingdom and France. TAT-8 was followed by TAT-9 in 1991 with a capacity of 80,000 simultaneous telephone calls. In the 6 years following, there has been a proliferation of undersea fiber optic cables providing a network of communication highways that

Figure 1-7   Dispersion in Optical Fibers.

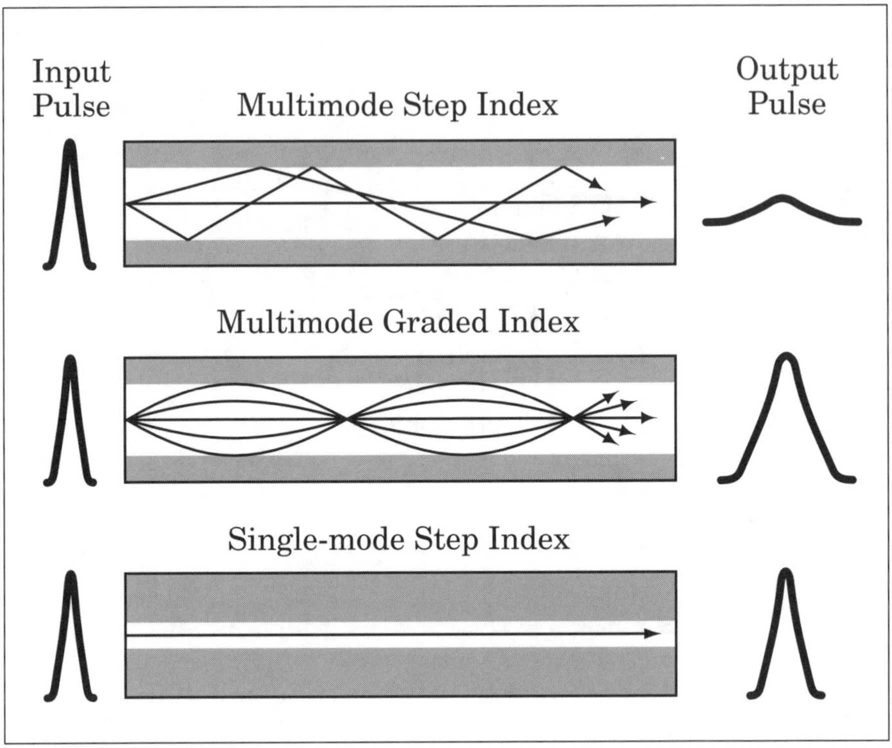

interconnect even the remotest parts of the world. These are discussed in more detail in Chapter 4.

In the early to mid-1990s the advances in fiber optic telecommunications technology accelerated. High-speed semiconductor logic was developed to switch digital circuits from 1s to 0s and back again at gigabit rates. Lasers, couplers, and filters were developed that would allow multiple light wavelengths at controlled bandwidth and spacing, each modulated with a different signal at rates up to 10 Gbps, to be coupled to one fiber at the sending end and separated at the receiving end. This technology is called dense wavelength division multiplexing (DWDM), and is similar to the concept of FDM used for wire line carrier and microwave radio, except that the light frequencies are a trillion times higher and the signals are lightwave instead of electrical. Transmission at a total rate of 1.1 terabits per second (Tbps), using wavelength division multiplexing (WDM) of 55 different wavelengths of light, each modulated at a rate of 20 Gbps has been demonstrated. Total transmission capacity of one fiber is the equivalent of 15 million voice frequency circuits.

At a transmission rate of 45 Mbps, which was the transmission rate for the first commercial telephony fiber optic systems, the information in one 30-volume set of the *Encyclopedia Britannica* can be transmitted in 1 second. At a transmission rate of 1 Tbps, information equivalent to 20,000 sets of the *Encyclopedia Britannica* can be transmitted in 1 second.

Modulated lasers capable of transmitting 110 analog television channels on a single fiber are available today. Products using these lasers are probably more suitable for CATV and interexchange trunk circuits than for distribution to subscribers.

Optical fiber amplifiers (OFA) have been developed that allow the light signal itself to be amplified rather than converting it to an electrical signal, regenerating it, and converting it back to optical form before retransmitting. Recent laboratory systems have demonstrated transmission at digital rates of 10 Gbps over repeaterless spans 500 km long with bit-error rates of 10^{-16} (not more than 1 bit in error out of 10 trillion bits). Experiments by AT&T personnel using fibers in existing undersea fiber optic cables in the Atlantic (TAT-12) and Pacific (TCP-5) oceans have achieved transmission rates up to 30 Gbps over distances of more than 16,000 km. Light signals of different wavelengths, each carrying digital data at many gigabits per second have been combined by WDM to transmit a combined rate of 1 Tbps over a single fiber.

New Network Technology

In addition to the advances in transmission technology that have increased the information carrying capacity and distance capability of transmission media, new networking technology that can improve the efficiency with which the capacity is utilized has been developed. The most notable improvement is packet switching, which is the basis for asynchronous transfer mode (ATM), also called "cell relay" (see Figure 1-8). ATM uses the transmission medium only when information is being sent, providing "bandwidth on demand" as an alternative to assigning FDM and TDM channels on a full-time basis and having idle channels and wasted bandwidth when information transfer is not required. ATM is specified in the IEEE Standard 802.6 for metropolitan area networks (MAN), switched multimegabit digital service (SMDS), and broadband integrated services digital network (BISDN) services.

Digital data compression is a technique that can reduce the bandwidth required for transmission of digital video for conferencing and motion pictures. Video frames are composed of many little dots, called pixels. Information about the brightness and color of each pixel must be transmitted sequentially as the picture is scanned. To reduce the number of bits per second required to transmit the video information when a new frame is scanned, bits are sent only to indicate pixels that have changed since the preceding frame. The background of the picture is constant if the camera is not moving, and if a person in the picture is moving only the lips, then only bits that represent changes in the lips must be sent. The resultant bandwidth reduction can be between one to two orders of magnitude.

Figure 1-8 ATM Cell.

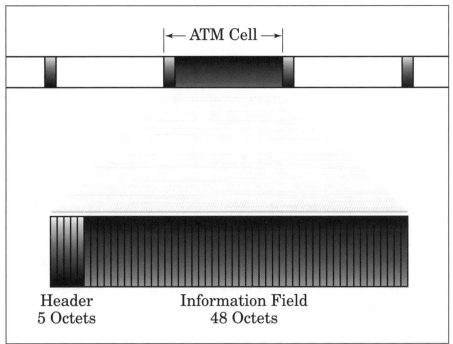

|← ATM Cell →|

Header   Information Field
5 Octets   48 Octets

Telecommunications Act of 1996

In February, 1996, the U.S. Congress passed the Telecommunications Act of 1996, which allows long distance carriers and CATV companies to provide long distance access, local telephone services, and data transmission services to subscribers in addition to television programs. It also allows the RBOCs to provide long distance service to its subscribers and to build video dial tone networks so that they can become common carriers to providers of television programming.

The impact will be open, active competition between established local telephone companies, CATV operators, long distance telephone companies, and new communications companies.

Interexchange carriers will move into local service, local telephone companies will move into long distance service and CATV service, and CATV providers will move into local service and competitive access.

The CATV companies have already placed orders for large quantities of modems that will provide high-speed access to the Information Superhighway. Modem vendors and the CATV companies have agreed to standardize all TV modem equipment. The CATV networks will provide data services at rates from 10 Mbps to 40 Mbps.

Also in 1996, SBC Communications (the Southwest Bell Company RBOC) purchased Pacific Telesis (another RBOC) while NYNEX and Bell Atlantic (two other RBOCs) merged, putting back together some of what Judge Greene rent asunder in his "Modified Final Judgment" of 1984.

Throughout just one and a half centuries of invention and improvement of electrical telecommunications technology, the world has advanced from the simple telegraph link from Baltimore to Washington to a proliferation of networks of wired and wireless communication channels and switching centers that can be internetworked to provide a seamless global network. Through this network, information in the form of printed and spoken words, pictures, and data can be transferred between most points in the world almost instantaneously. The spectrum of frequencies utilized to provide these communication channels is presented in Figure 1-9.

The idea of a national information infrastructure (NII), developed in the United States during the Clinton administration, and the agreement between 68 countries of the World Trade Organization to allow telecommunications organizations of other countries to provide services within their boundaries will further the realization of an Information Superhighway. This

Figure 1-9   Telecommunications Frequency Spectrum.

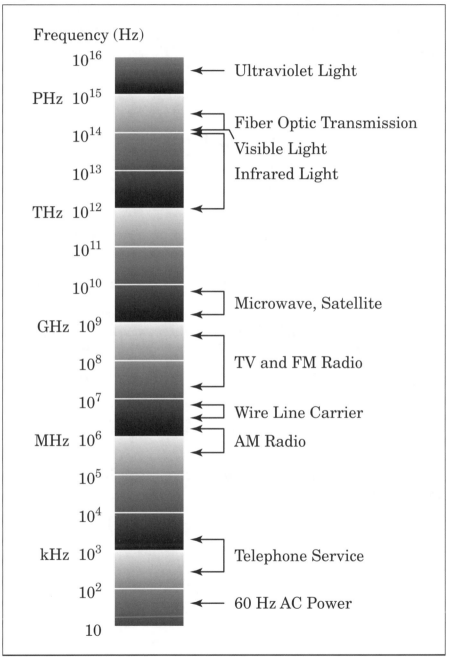

Information Superhighway will consist of a myriad of information sources and databases as well as the internetworked telecommunications systems that will provide the means of access for users.

Bibliography

Blyth, John W., and Blyth, Mary M., *Telecommunications: Concepts, Development, and Management,* Mission Hills CA: Glencoe Publishing Company, 1985.

International Telecommunications Union, *From Semaphore to Satellite,* Geneve, ITU, 1965.

Martin, James, *Telecommunications and the Computer,* 2nd ed., Englewood Cliffs, NJ: Prentice-Hall, Inc., 1976.

Nellist, John G., *Understanding Telecommunications and Lightwave Systems,* New York, NY: IEEE Press, 1996.

Oxford English Dictionary, 2nd ed., Oxford: Clarencon Press, 1989.

Reference Data for Radio Engineers, 6th ed., New York, NY: Howard W. Sams & Co., Inc., 1985.

Webster's Third New International Dictionary, Springfield, MA: Merriam-Webster, Inc., 1981.

2 The Computer and the Internet

I think there is a world market for about five computers.

— Thomas J. Watson, IBM chair, 1943.

A Brief History

The development of the computer was certainly one of the most important events of this century. On June 21, 1948, the first stored program in a general-purpose computer, written by Tom Kilburn, ran successfully at the University of Manchester in Britain on a computer nicknamed "Baby." The event marked the beginning of the age of information technology.

It took another 30 years for this electronic brain of the 1940s to blossom into the huge computer industry of today. By the late 1970s, most of today's leading players were in position to create the modern information technology industry.

The first computer was developed by Bell Labs in 1940 using electromechanical relays. The first mass-produced electronic computer was the UNIVAC built in 1951 by Remington Rand Corporation. In 1964 International Business Machines (IBM) Corporation introduced the System 360 line of mainframe computers. The first personal computers were introduced in 1977 by Apple, Radio Shack, and Commodore. More than 5 million of these machines were eventually sold.

After the Second World War, most of the components needed to build the modern digital computer were in place. Only one thing was missing, a viable way of storing and retrieving data, that is, an electronic memory. Without a memory, computers were made to work by a complicated and time-consuming process of setting switches. As a result of research to improve cathode ray tubes (CRT), a method of using the CRT to make a digital memory was developed. The CRT's tendency to hold onto an image, while not welcome on a radar screen, could be used to store digital information, and retrieve it. It held data in a way that could be accessed directly, a true random access memory (RAM).

Computer Miniaturization

Scientists at Bell Laboratories in the United States introduced the transistor in 1947. The transistor is a solid-state device and does not require a heated cathode like the vacuum tube. By 1959 Texas Instruments and Fairchild Semiconductor successfully produced integrated circuits with transistors, capacitors, and resistors placed on a square of silicon.

Integrated circuits (ICs) containing millions of electronic components have evolved over the past three decades and today form a significant part of the information technology industry. A steady stream of products based on ICs has been developed for use in telecommunications, automobiles, appliances, homes, business, and entertainment systems. Microprocessors are the brains of the modern computer. Today's processor technology is classified as Micro Age technology. Soon we will be moving into the Nano Age. The emerging nanotechnology will have a major impact on the price/performance of current and yet-to-be discovered products and services.

Microprocessor Size

A processor contains millions of electronic components. For example, IBM's P2SC processor contains 15 million transistors. The size of this IC is smaller than the human thumb nail. The evolution of processor size is shown in Table 2-1.

Table 2-1   Microprocessor Size.

Year	Size in Microns	Size in Nanometers
1985	1.00	1,000
1990	0.50	500
1995	0.35	350
2000	0.20	200
2005	0.10	100

Microprocessor Speed

Processor clock speed is an important specification of today's computer. The higher the speed, the better the performance of your personal computer (PC). For example, an incandescent light bulb flickers 60 times in 1 second, while Intel's Pentium II processor cycles through 450 million times in 1 second. Processor speed has evolved along with the ongoing reduction in size of the semiconductor process technology. Table 2-2 shows clock speeds for a number of commercial processors.

From Megabytes to Gigabytes and Beyond

Miniaturization has not been confined to just microprocessors in the computer. It has made computers much faster and cheaper, but it has also allowed the storage of huge quantities of information in a small space.

Table 2-2   Microprocessor Speed.

Vendor	Product	Clock Speed in MHz
IBM	P2SC	135
HP	PA-8000	180
SUN	ULTRASPARC	250
Intel	Pentium II	450
Digital	21164	500

In 1956 IBM introduced the first commercial hard disk drive, which was about the size of two refrigerators with 50 disks. Each disk was 24 inches in diameter and contained 5 megabytes (MB) of memory. The cost was about $10,000 per megabyte.

In 1963 the first disk drive with a removable disk pack was introduced. By 1976 nonremovable Winchester disk drives led to a significant drop in costs, about $100 per megabyte. Laboratory demonstrations in 1989 previewed a 1-GB/in^2 recording density in a magnetic medium. The arrival in 1997 of a 8.4-GB hard drive broke the gigabyte barrier and lowered the cost of computing to only 25 cents per megabyte. In 1998, IBM announced it will offer a 25-GB hard drive for personal computers. The new drive has 5,000 times the capacity of their first drive in 1956.

The computer power and memory that can now be concentrated in the volume of a sugar cube would, in 1950, have required a trillion sugar cubes. One thing that hasn't changed since the 1950s is the physical way in which information is stored.

Digital computers understand only binary numbers, which are represented by combinations of the two digits 1 and 0. Each 1 and 0 is called a bit (binary digit). Codes have been developed to translate letters (A - Z), numbers (0 - 9), and punctuation marks into binary 1s and 0s that computers understand. Each character is represented by a unique combination of a fixed number of bits, usually from 6 to 8, depending upon the particular code that is used (see Figure 2-1).

If you take any magnet, it generates a magnetic field that has an effect on other magnets or metal. The magnetic field points in some direction. A single computer memory element is just a magnet designed so that this field points either up or down. Inside the computer, this memory element then stores 1 bit of information; it is either on or off (the number 1 or 0) depending on whether the field points up or down. The big difference between now and 1950 is in the size of these magnetic memory elements. In 1950 each one was about the size of a sugar cube. Now they are far too small to see and there are many billions of them on the surface of the hard disk drive in any personal computer. In order to continue this miniaturization process, scientists have been pursuing another area of investigation called *molecular nanotechnology*. Even though the existing

Figure 2-1   Letter "C."

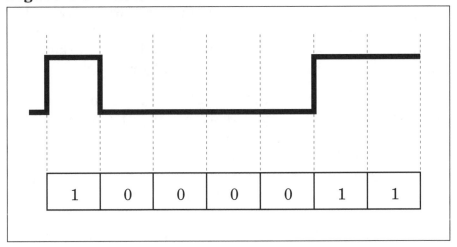

memory elements are small compared to us, they are still huge compared to atoms. And each atom is also a tiny magnet, which can be arranged to point either up or down. At the rate we are going, by 2050 the amount of information we will be able to store in the volume of a sugar cube may be greater than that in all the computers in the world today.

The incredibly rapid evolution of computing power, as well as changes in communication and information technology, have made most predictions about the future very suspect. The evolution of the Internet in just the last 5 years is an example of how the future impact of a new technology can be disproved very quickly.

The Millennium Bug

The millennium bug (or Y2K, as the computer industry has dubbed it) is the doomsday scenario in which almost every computer made before 1995 will go back 100 years to 1900 instead of clicking over into the year 2000. Many software experts believe that millions of computer systems will shut down on Saturday, January 1, 2000, creating a massive power blackout because computer-controlled power generating stations will shut down. Telephone service will fail, banks and supermarkets will close, and all commercial aircraft will be grounded.

The bug at the center of the year 2000 crisis is fairly simple. The early mainframe computers were very primitive by today's standards. These machines were short on expensive main memory, and the transfer from tape to memory was slow. Memory space was at a premium and it was the job of programmers to make the most of it. The system programmers set aside only two digits to denote the year, as in 07/20/64 rather than 07/20/1964. At the time, the shortcut saved a significant amount of space and became standard practice. Programmers were not concerned about any potential problem that might develop in the future. But if knowing how to fix the problem is easy, finding what to fix is not. Problems can arise for several reasons, but the most common is that code, particularly older code, is undocumented. The programmers who wrote it and those who later patched it did not leave an annotated version behind explaining how and where they used date information.

Another common problem is that the original source code has been lost. Only the working program remains, a seemingly endless string of 1s and 0s that must be interpreted, or reverse-engineered to see what the original code looked like.

Fortunately, the majority of firms are taking action to make sure this scenario will not come to pass, at a collective cost estimated at $50 to $600 billion.

The millennium bug has certainly proved that the world has come to rely on computers as an important part of everyday life.

> *There is no reason for any individual to have a computer in their home.*
>
> — Ken Olson, president, Digital Equipment, 1977.

The Internet

Until something better comes along, today the Internet (also called the Net) is the closest thing we have to an Information Superhighway.

The Internet actually originated in 1969 as an experimental network by the Advanced Research Projects Agency (ARPA) of the U.S. Department of Defense. The network (originally called ARPANET) was designed to enable scientists to communicate among themselves. ARPANET originally consisted of four computers, but by 1972, 50 universities and military research sites had ARPANET access.

During the 1980s, several other networks (including a National Science Foundation network of five supercomputers) sprang up. Eventually, all of these public and private networks were interconnected to enable any computer on one of the subnetworks to access computers anywhere in the entire network.

The community of Internet users, currently estimated at more than 100 million, is expected to surpass 700 million by the end of this century. This unprecedented growth is the exponential value of networking, articulated by Metcalfe's Law. Named for Robert Metcalfe, the father of the Ethernet and a co-founder of 3Com Corporation, Metcalfe's Law states that when you connect any number (n) of machines, you expand the potential value of those machines by an exponential factor (n^2). It doesn't matter whether the machines are cars, telephones, or computers; cars without roads and computers without networks are of limited use. Provide the infrastructure and their value grows exponentially.

Cyberspace

The term *cyberspace* was first used by a science fiction writer, William Gibson, a young expatriate American living in Vancouver, Canada. By 1989 it had been borrowed by the online community to describe today's increasingly interconnected computer systems, in particular the millions of computers plugged into the Internet. The Internet, of course, is more than a telephone call. It encompasses the millions of personal computers connected by modems via the telephone system to commercial online services. Some of the major online Internet service providers (ISPs) are America Online and CompuServe. They must compete, however, with hundreds of direct Internet access providers that offer cut rate Internet-only access.

One factor fueling the Internet's remarkable growth is its grass-roots structure. Most conventional computer systems are hierarchical and proprietary; they run on copyright software in a pyramid structure that gives dictatorial powers to the system operators who sit on top. The Internet, by contrast, is open (nonproprietary) and very democratic. No one owns it. No single organization controls it. It crosses national boundaries and answers to no sovereign.

The Internet is far from perfect. Largely unedited, its content is often tasteless, foolish, and uninteresting. It can be very habit-forming and an enormous waste of time.

The Internet, however, is changing rapidly. A lot of the development efforts have shifted from the more traditional uses of e-mail and newsgroups to the more passive and consumer-oriented "home pages" of the World Wide Web (WWW), a system of links that simplifies the task of navigating among the many offerings on the Internet.

Web Browsers

You can't surf the Internet without software, and new software is making the surfing much easier.

For years, working on the Internet meant working in text, typing in long and often complicated command codes to get where you wanted to go. The World Wide Web changed all that. Today, the Web is a constantly evolving place where nothing stands still for long.

There are many Web browsers available, but only two have captured the majority of Net surfers. They are Microsoft Explorer and Netscape Navigator. Explorer and Navigator run programs written in Java. Java is a new programming language developed by Sun Microsystems Inc. Most leading computer platforms already feature built-in Java support. Nearly every major software developer has licensed Java to develop new applications or rewrite existing ones.

The expected flood of commercial Java applications is already appearing, led by Corel's WordPerfect and accompanying office suite.

Initial demand for Java was to provide interactivity and animation to World Wide Web pages through the use of specialized applications or "applets." Corporate organizations quickly began to recognize the potential of Java, particularly the ability to deliver computer applications to users across an in-house network, or intranet, or even across the Internet itself. In many cases, Java has eliminated the need for a traditional personal computer with its never-ending cycle of software and hardware upgrades, administration, and maintenance. It has generated interest in the network computer (NC). Oracle Corp. originally referred to this visionary NC as the "under-$500 computer," which would provide everyone with low-cost access to the Internet.

The NC is essentially a personal computer with fewer moving parts, no disk drives, and relying on the network server to provide the processing power, storage, content, and administration, all through one language, such as Java. However, with the recent drop in computer prices this concept has stalled.

But Java itself is platform-independent. It allows applications to run on any device, including mainframes, PCs, NCs, cellular phones, or even household appliances. This means software developers can create one version of any application and deploy it on a wide array of systems.

Marketing on the Internet

Marketing on the Internet is already big business. Companies are scrambling to get on. The Net is an excellent marketing tool. It can convey a lot of information and can easily build a profile of visitors allowing one-on-one marketing on return visits.

Internet advertising is a relative bargain. The price of a banner on the most frequently visited home pages ranges from $5,000 to $20,000 a month, compared with $15,000 for a half-page, full-color advertisement in the leading computer magazines. This has enormous appeal for publicity hungry companies with limited advertising budgets.

Even more impressive are estimates that the Internet revenues will explode to $195 billion by 2000, from $21.7 billion in 1998. The amount of business-to-business transactions, for example, is estimated to soar to $175 billion from $15.6 billion, while retail sales will hit $20 billion from $6.1 billion.

Internet advertising revenue is also concentrated: 30% of the total is done by the 10 leading home pages. This list includes Playboy, ESPN, Netscape, Pathfinder, Hot Wired, and Yahoo!

The more the Internet becomes a legitimate advertising vehicle, the more it justifies the existence of companies that sustain themselves by selling space on their home pages. One of the highest-profile companies to implement this economic model is Sunnyvale, Californian-based Yahoo! Inc. It is now the world's most popular Web site.

The company was founded in 1994 by two graduate students,

Jerry Yang and David Fido, at Stanford University. Yang, whose business card describes him as "chief yahoo," has been able to get such major companies as NBC, CitiBank, and Colgate-Palmolive on board. Yahoo! has also become the de facto search vehicle for the Internet because it offers "surfers" an easy-to-use interface to hunt for information.

It caters to the interests of nearly everyone by providing jumping off points to subjects such as business, entertainment, technology, and economics. From there, Yahoo!'s search engine software provides people with access to its huge database of 115 million Web pages, which it collects by constantly scouring the Internet. In a sense, Yahoo! has popularized the Internet by putting information at people's fingertips and eliminating the apprehension many new "surfers" have when they launch themselves on the Internet. It averages 31 million "visitors" per month.

The number two search engine, Excite, draws more than 16 million users per month. While thousands of companies promote their products on the Internet, many have begun selling online. The Internet business boom is divided into five sectors: hardware, software, Internet service providers, retail services, and content.

Hardware

The major players in this sector are Cisco Systems, 3Com, Sun Microsystems, and IBM. The hardware sector is the most vibrant in the Internet "economy" sector as access providers, telecommunications firms, and major corporations build their networks. Players like Cisco Systems and 3Com are making significant revenue and profit by selling equipment to manage traffic, while IBM and Sun Microsystems are providing computers and servers. The long-term outlook is promising, particularly as corporations build intranets, the internal communications networks that use Internet-based tools. Cisco Systems' site on the World Wide Web has generated sales of $2 billion a year, or nearly one-third of its total sales.

Predictably, high-technology companies have taken an early lead in Internet commerce. These companies are already wired to the Internet, as are their customers. Dell Computer Corporation, for example, began selling personal computers on their Web site in 1996, and its Internet sales are now almost $5 million per day.

Software

The key battle over software is between Microsoft and Netscape to win over customers to their Web browsers and applications that allow electronic commerce. A growing number of software developers are also cashing in, including Toronto-based EveryWare Development Corp., which makes tools to create dynamic Web sites. As the Internet becomes more popular, software developers should continue to make sales and profits. The most explosive area is expected to be server software that will allow companies to navigate, collaborate, and do secure transactions over the Internet.

Internet Service Providers

America Online (AOL) acquired rival CompuServe in 1997, which it now operates as a separate service. AOL is now the number one Internet service provider with over 12 million members. Most access providers enjoy growing revenues, but the cost of building infrastructure and attracting new customers makes profit unlikely, for now.

Retail Services

Although electronic commerce attracts a lot of attention, it has failed to catch on with consumers who are worried about security when sending credit card numbers over the Internet. There are success stories, such as Seattle-based book seller Amazon.com, which was started in 1995.   Customers can choose the title, type in their address and credit card number, and the warehouse will mail it out, usually in 1 or 2 days, and often at a discount.   Four years after its launch, Amazon.com has 2.5 million customers worldwide, and sales of $607 million.

Eddie Bauer, the outdoor clothing retailer, has an online operation that has been profitable since 1997, and it is growing at a rate 300% to 500% a year.

For some companies, the Internet revolution has been very unpleasant. The 230-year-old *Encyclopedia Britannica* laid off its entire home sales force in North America as a result of the Internet. An inexpensive CD-ROM has replaced the 30-volume set of books worth $1,250.

While retail sales are growing rapidly, the Internet still only represents no more than 1% of the United States' $8.5 trillion economy.

Equity Trading

All across North America, full-service, discount, and strictly online brokerage firms are carving out a niche for themselves in the highly competitive World Wide Web. In the United States, online investing is forecast to surge to 4.9 million accounts by the end of 1998, from 2.7 million in 1997. It has been predicted that 12.5 million accounts will be traded online by 2002. In Canada, Green Line Investor Services will have

100,000 clients registered to trade on the Web by the end of 1998. While the stock prices of Internet companies have been climbing like a rocket, it is estimated that only 3 of about 85 companies are actually making a profit. They are Yahoo!, Mindspring Enterprises, and America Online.

Content

Given the Internet's ability to reach millions of potential customers without high distribution costs, it has become popular with magazine, newspaper, and newsletter publishers looking to extend their franchises or create online entities from scratch. Unfortunately, Web surfers are reluctant to pay subscription fees for content, so advertising has become the primary revenue source. With few exceptions, content providers are spending a lot of money, but they will continue to post losses until the amount of advertising grows. *USA Today*, for example, was forced to abandon its subscription fee model, while Time Warner Inc.'s Pathfinder site, which features magazines such as *Time, Life*, and *Sports Illustrated*, has yet to proceed with plans to charge for access. A typical Web page is shown in Figure 2-2.

While Internet commerce is in its early days, now is the time for business to experiment with possibilities. A company that fails to adapt to the Internet is in danger of either losing its competitive position or missing a vital new opportunity. Most businesses are very positive about the Internet, despite the enormous hype the World Wide Web has received recently.

Electronic Transactions

In July 1997, MasterCard conducted the first end-to-end transaction on the Internet using the Secure Electronic Transactions

Figure 2-2   Typical Web Page.

(SET) protocol. Developed by Visa and MasterCard, SET is a standard protocol for safeguarding payment card purchases made over the Internet.

SET is designed to operate both in real time, as on the World Wide Web, and in a store-and-forward environment, such as e-mail. As an open standard, it is also designed to permit consumer, merchant, and banking software companies to develop software independently for their respective clienteles and to have them interoperate successfully.

SET assumes the existence of a hierarchy of certificate authorities that vouch for the binding between a user and a public key. Consumers, merchants, and acquirers must exchange certificates before a party can know what public key to employ to encrypt a message for a particular correspondent (Figure 2-3).

The operation of the SET protocol relies on a sequence of messages. In the first two, the consumer and merchant signal their intention to do business and then exchange certificates and establish a transaction ID number. In the third step, the consumer purchase request is negotiated outside the protocol. This request is accompanied by the consumer's credit card information, encrypted so that only the merchant's acquiring bank can read it. At this point, the merchant can acknowledge the order to the customer, seeking authorization later (steps five and six) or perform steps five and six first and confirm authorization in step four. Steps seven and eight give the consumer a query capability, while the merchant uses steps nine and ten to submit authorizations for capture and settlement.

Education

In the field of education, Canada leads the world in bringing the world to its classrooms. Developed to introduce the Internet to Canadian schools, SchoolNet was launched as a government/private-sector initiative in August 1993, when the Internet was in its relative infancy. The aim of SchoolNet's pilot phase was to connect 300 schools. But after an overwhelmingly successful first year, the goal now is to hook up all 16,500 elementary and secondary schools in Canada with all of the country's libraries and universities.

Integrating the Internet into the classroom empowers learners with the electronic networking skills that they require for the increasingly competitive information-based job market of the 21st century.

Students from across the United States and Canada can access the Internet 24 hours a day, 7 days a week. They have

Figure 2-3   Secure Electronic Transactions (SET).

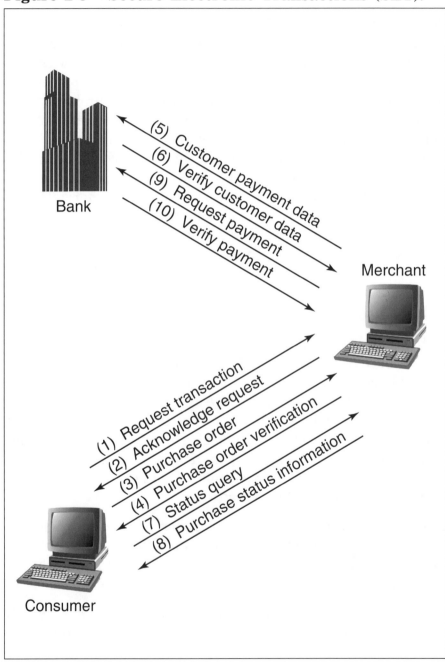

user-friendly, high-speed access to world-class libraries and art galleries, 400 scientists, engineers, and advisors from around the world, electronic versions of newspapers and magazines, and dozens of electronic discussion groups created just for students. They can also work on multimedia projects with classmates, talk to kids around the world, and even develop their own Web page.

Internet Access

Internet access from the home and many businesses presently takes place at speeds of 14.4 and 28.8 Kbps, using relatively cheap and reliable modems that operate over existing voice grade copper pairs (Figure 2-4). Such speeds are adequate for e-mail and small file transfers but result in lengthy delays for graphics, such as those used in Web pages.

Effective use of the Web as a business, entertainment, and educational tool will require more bandwidth than is achievable over voice grade lines. New technology now being deployed will increase this bandwidth.

The telephone companies are offering the integrated services digital network (ISDN), which can provide access at up to 128 Kbps. This is fast enough for low-quality video, such as desktop videoconferencing, but is still too slow for video, even at VCR-quality levels.

Another option offered by the telephone companies is a broadband modem called asymmetric digital subscriber line (ADSL), which operates over existing copper pairs at up to 6 Mbps. The upstream rate is 640 Kbps. The limitations of twisted pair copper wire restrict it to about 12,000 ft at 3 Mbps and 8,000 ft at 6 Mbps, so some telephone company plant upgrading will be necessary to accommodate some customers.

Figure 2-4   Internet Access.

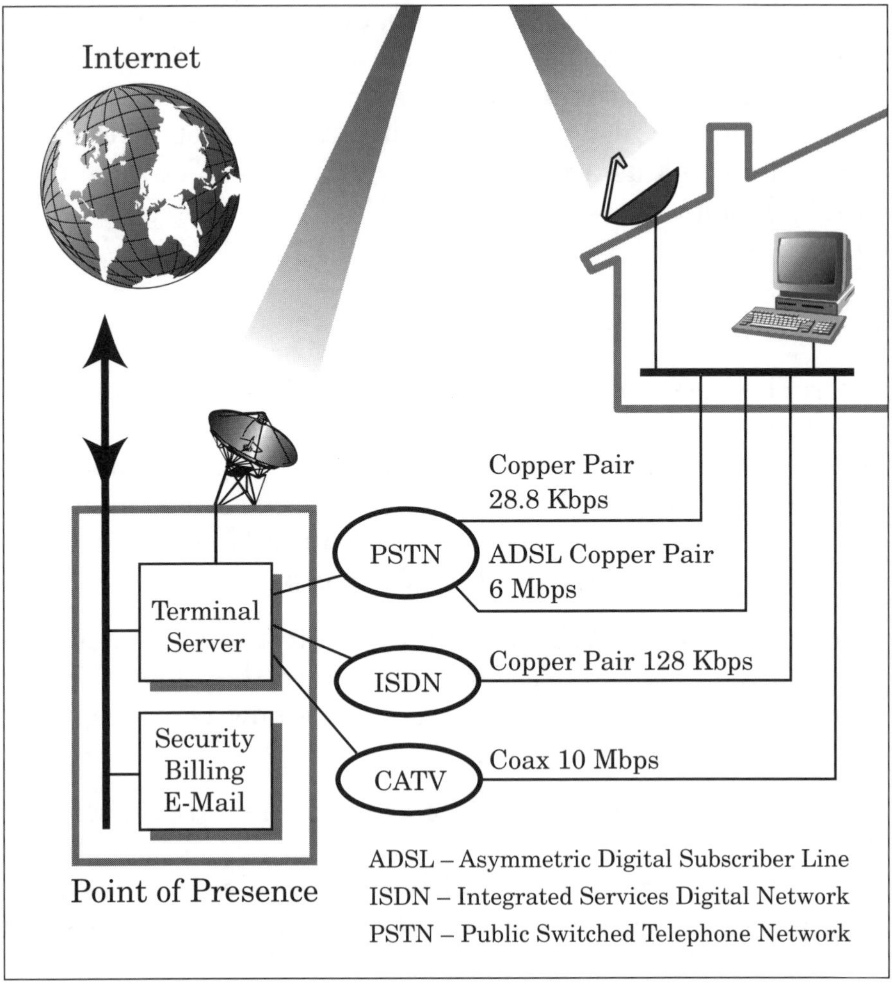

However, about 50% of homes in the United States are within this distance.

The cablevision operators are using their wide bandwidth coax cable to provide high speed Internet access using cable modems. These devices operate in a similar fashion to telephone modems except at a higher speed, usually on the order of 10 Mbps, but in some cases up to 27 Mbps. There are two problems with cable modem technology. First, the bandwidth is shared, which means that not all of the 10 to 27 Mbps are available at all times; and second, the current coax system architecture, which was not designed for two-way communication, is noisy in the upstream direction. The drop in capacity will be felt when it is shared by 100 or more subscribers.

The direct broadcast satellite (DBS) operators are now offering Internet access. Hughes Network Systems is providing Direct-PC, which utilizes a DBS-type dish slightly larger than that used for DirecTV (21 inches versus 18 inches) with different frequencies and hardware. At present, the system is unidirectional (users can download files and view Web pages), but upstream traffic must take another route. Speeds are up to 12 Mbps and charges are based on the number of megabytes of data per month.

On-Ramp Congestion

Some experts are predicting that there will be as many as 700 million Internet users worldwide by the year 2000. The explosive growth of Internet use is intermittently clogging phone networks, making it harder for people to complete calls during busy hours.

Phone companies in the United States and Canada have spent billions of dollars laying new fiber optic cable to increase the capacity of their systems. Then along came the Internet, which gives people access to endless amounts of information, and gives them the opportunity to spend endless hours exploring online. The Net's rising popularity has left phone companies to deal with increased call volumes and starkly different call patterns. This has changed the daily peak times for the phone systems.

Three years ago, the network had to contend with one peak at 10:30 am and another one at 3 pm. Nowadays, the systems face their greatest demand at 10:30 pm. Worse still, this new peak is about 50% larger than the old ones, mainly due to Internet traffic. It's not the sheer volume of Internet calls at these times of the day that is causing headaches for the carriers. Most companies already have high-capacity fiber optic cable linking their central office switches. The congestion is caused by the increasing length of time users stay on the phone, known as hold time.

Unlike voice callers, Internet users often leave the connection open for hours once they've dialed into their ISPs and begun to surf. These calls generate money for the ISP, but nothing for the phone companies.

Also, many ISPs now charge a flat-rate monthly fee for Internet access instead of charging by the hour. So customers are no longer discouraged from surfing the Net for as long as they want.

In 1995 customers spent an average of 3 minutes making each phone call.

By 1997 that average had risen to 20 minutes. With longer hold times now commonplace, the calling load on the networks at peak times of the day can cause problems.

A typical switch can handle as many as 2 million calls in an hour if these calls average 3 minutes. But if you boost the average hold time to 20 minutes, that switch can handle only 300,000 calls hourly. As a result, switches can get congested as telephone and Internet calls fight for space on the network.

Telephone companies are considering two possible solutions. They can install a new Internet product that identifies ISP phone numbers and separates that traffic from the voice signals, or they can use ADSL, which lets users access the Internet at faster speeds. Either way, the carriers hope to remove heavy Internet users from the voice side of the phone network. Once the ADSL calls reach the central office switch, they can transport the traffic by ATM, a technology designed specifically to transport data signals.

In the end, the wider on-ramps will enable traditional phone users and Internet surfers to access the Information Superhighway without any traffic congestion.

Conclusion

The Internet is growing faster than all other technologies that preceded it. Radio existed for 38 years before it had 50 million listeners, and television took 13 years to reach that mark. The Internet crossed that line in just 4 years.

The reason for the Internet's incredible impact goes far beyond the simple fact of unprecedented growth. One explanation lies in the fact that the Internet has added a remarkable new dimension to communication. It has put the power of information into the hands of the consumer by making it easily accessible. Where do you go to find information about a potential holiday spot? A company you're interested in dealing with? An airline schedule? The most up-to-date results of the U.S. presidential election?

Today, most computer users head directly to the Internet. It is global, more convenient, more current, and more powerful than any other method.

For the average person, the Internet has many limitations, the main one being bandwidth—the size of the pipe going into the home. That pipeline, which carries all the information and entertainment to the home or business, must also have sufficient bandwidth to carry all the upstream data from the home to the network.

As a result, today's Internet is just the first step toward the Information Superhighway of the future. Its construction has already begun by the telecommunication, cablevision, and satellite companies around the world.

They are rebuilding and upgrading their local networks with fiber optic cables that will provide the bandwidth to carry all the new services. In addition, satellite operators are building networks that will allow consumers to link home computers to the Internet via satellite. Some of the two-way satellite services will package satellite TV and Internet access on a single dish.

The following chapters will describe the technologies that are being used to build the global Information Superhighway.

Bibliography

Comerford, Richard, and Perry, Tekla S., "Brooding on the Year 2000," *IEEE Spectrum,* June 1998, p. 68.

Fowler, Thomas B., "Internet Access and Pricing: Sorting Out the Options," *Telecommunications*, February 1997, p. 41.

Sirbu, Marvin A., "Credit and Debit on the Internet," *IEEE Spectrum,* February 1997, p. 23.

Smyth, George C., "The Internet: Powering the Information Age," *Nortel Networks Telesis,* No. 102, December 1996, p. 3.

3 Who Uses the Information Superhighway?

Horses are here to stay. The automobile is only a novelty, a fad.

> — The president, Michigan Savings Bank, 1903,
> advising Henry Ford's lawyer not to invest
> in the Ford Motor Co.

A Brief History

Much of the planned seamless global telecommunications network that we have begun to call the Information Superhighway is in place now and almost all of us use some part of it every day. Plain old telephone service (POTS) is something we all use to call within buildings or across town or perhaps to neighboring towns that are in our local calling area. (You may call telephone numbers with prefixes defined by the local service provider as part of your local calling area without paying toll charges.) The communications facilities that provide POTS are one part of the Superhighway.

Since the deregulation of AT&T, POTS services have been provided by one of the Regional Bell operating companies (RBOCs), one of the GTE telephone companies, or one of the several hundred independent telephone companies. The Telecommunications Act of 1996 allows long distance carriers and CATV companies to provide local telephone and data transmission services to subscribers so that many new facilities are expected to be added to the local access part of the Superhighway. Local distribution and access are discussed in Chapter 7.

When we call other cities we extend our use of the Super-highway to long distance networks in addition to local access. Prior to deregulation in 1982, long distance service within the United States was provided by AT&T. Since deregulation, it has been provided by special common carriers like Sprint, MCI, and AT&T. The Telecommunications Act of 1996 allows the telephone operating companies and the cable TV companies to compete for this business. In other countries in the world, net-works are in place and internetworking technology has been developed to implement seamless international communica-tions so that international calls can be direct-dialed and connections made almost instantaneously.

So, what does the world do with the tremendous bandwidth already installed and increasing daily in the fiber optic cables, satellites, cellular radios, and hybrid fiber/coax local distribu-tion systems? What types of services will travel over the Superhighway? What are the current applications for the ser-vices, and what applications do we see for the future?

Types of Communications Services

The ITU-T Study Group 18 (SGXVIII) has classified broadband communications services that could be supported by B-ISDN, which will be part of the Superhighway. Their classifications may be applied to telecommunications in general and are use-ful for contemplating the types of information that will flow over the Superhighway and the telecommunications media and technology that will be required to transport it.

The ITU-T considers two broad classifications: *interactive ser-vices* and *distribution services*. Interactive services provide two-way information exchange between points on a network, such as communications between two subscribers or between a subscriber and a database. Most of the information transfer consummated through the Internet is supported by interactive services.

Interactive Services

The interactive service classification is further broken down into conversational, messaging, and retrieval subclassifications. Conversational services provide real-time, bidirectional communications between users. Examples of conversational services that can be supported by the Superhighway include video telephone, video/audio information service, videoconferencing, POTS voice conversation between subscribers, and interactive data transmission between a user and a computer. Videoconferencing may be either point-to-point with one or more persons on camera at each end, or multipoint, where a group of people in different locations can be tied together in a conference connection and see, as well as hear, each of the members of the group.

In a typical multipoint videoconferencing arrangement, the video screen is split into multiple segments, one for each location. The video camera can be aimed at the entire group at a location and can zoom in on one person in the group when he or she is speaking.

Conversational high-speed facsimile exchange, high-volume file transfer, high-resolution image transmission, document exchange, and interconnection of local area networks also fall into the conversational classification.

Messaging services provide store-and-forward communications, such as e-mail, between users. This type of service may be expanded to include video mail, which would allow transfer of video clips or the contents of video cassettes, and document mail for the transmission of documents containing text and/or graphics.

Retrieval services allow users to retrieve information stored in databases or information centers. The type of information stored in the databases includes sound, text, graphics, images, and moving pictures.

The Internet is a fine example of an interactive service and it encompasses all of the subclassifications of interactive services: it provides conversational services in the form of voice conversations between users; it provides store-and-forward e-mail service; and it provides retrieval services when World Wide Web pages and databases around the world are accessed for information.

Distribution Services

The distribution services classification is divided into two subclassifications: distribution services without user individual presentation control, which are basically broadcast services such as broadcast television and radio, cable TV, and satellite TV; and distribution services with user individual presentation control, which allow the user to select the type of information of interest and control the start and order of the presentation.

Examples of services with user individual presentation control are dial-in information services and Teletext. Dial-in information services provide information on topics such as family medicine and bankruptcy as well as world, national, and sports news, and weather forecasts. Topics can be selected by pressing four-digit codes, and the messages can be interrupted, repeated, or backed up by keypad commands. The calls are free and there is no toll charge if the service is within your calling area. However, you might have to listen to a short commercial.

Teletext, which transmits a series of pages cyclically, is another example. By keying in command numbers on a decoder, teletext allows at the receiving end selection of pages to be presented on a television screen. The selected pages are decoded and stored by the decoder for display by the television.

Each of the types of services discussed above imposes a set of requirements on the media that is used to transport them. The Superhighway will need to provide media that will meet the requirements of all of the types. Much of today's infrastructure exists in the terrestrial and undersea fiber optic cables and terrestrial and satellite microwave radio networks of the service providers and private networks. This infrastructure is presently being expanded and will undoubtedly need continued expansion to meet the increasing use of telecommunications, especially considering that data communications are growing at the rate of 30% a year. Many new services, some which are presently beyond our imaginations, are expected to evolve in the future. A conceptual diagram of a residence equipped for access to the Information Superhighway is shown in Figure 3-1.

The Internet

Many consider the Internet to be the Information Superhighway. It is a vast global collection of networks that can provide access to worldwide information sources. Users access the Internet through local service providers to reach content providers, such as databases and news and information sources. The backbone connections to the many networks of the Inter-

Figure 3-1   Information Age Residence.

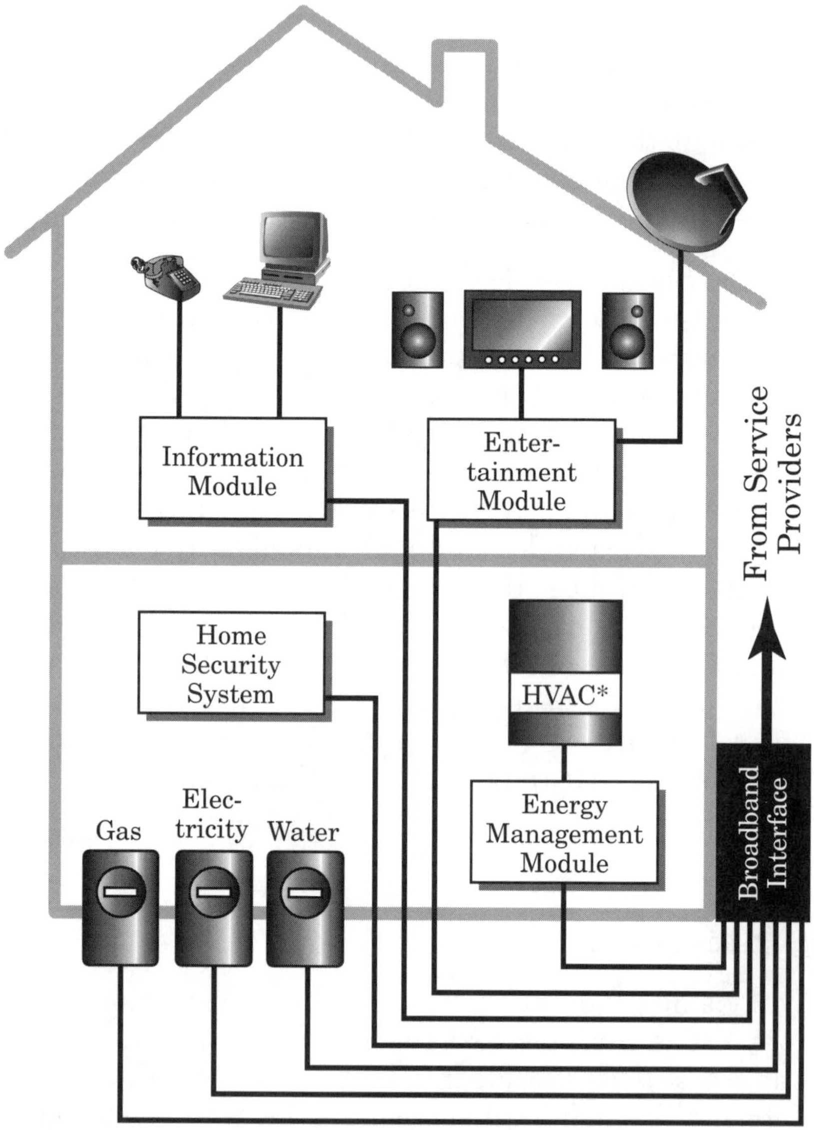

* Heating, Ventilation, Air Conditioning.

net are made through the infrastructure of the long distance service providers.

So, the Internet may evolve to be the Superhighway or may simply remain part of it. What is important is the evolution of a seamless network that can provide a pathway that will satisfy the requirements of all of the telecommunications applications that now exist and those projected for the future.

Some of the applications use Internet pathways; many do not, rather, they use the transmission facilities of the telecommunications service providers or those of their own private networks directly. Some of these applications will be discussed in the following paragraphs.

Business and the Internet

Business and commerce may be one of the largest applications of the Superhighway, and much of the use is through the Internet. The World Wide Web is one of the fastest growing uses of the Internet, with Web page growth estimated at 1,000% a year. The Web is a medium that is an advertising instrument, an information distribution agent, a means of generating sales leads, and an organization promotion vehicle. Companies that advertise on Web pages generate revenue through on-line orders for products and also realize savings from operations. For instance, requests for mailed information can be reduced considerably by allowing the information to be downloaded directly from the Web site. Other companies have found that the Web site can reduce the number of incoming toll-free 800 calls. Requests for pricing can go directly to an agent who can answer the request by e-mail.

By putting customer service, customer support, and product documentation (such as product application information) on a

Web site, customer satisfaction can be enhanced and the number of support calls, facsimile messages, and printing and postage costs can be reduced.

Another benefit of the Internet to business is the availability of enormous amounts of information stored in databases that can be accessed.

Businesses also use the Internet to interconnect geographically dispersed local area networks. Many companies are already using World Wide Web servers for data communications applications, and, by using ISDN services, companies can access the Internet at four to five times the digital rate of the fastest modems.

Companies are using Internet technology to build intranets, which are accessible only to their customers or their employees. Intranets provide subscribers with specialized information (prices, historical information, analytical data, and research information of stocks and bonds for example), for a monthly fee, and can be connected directly into the customers' own intranets. At the end of 1996, 64% of the top 1,000 companies in the United States had intranets.

Computer-Telephony Integration

Computer-telephony integration (CTI), which links telecommunications with databases to create single integrated business information systems, has been used for several years in call centers to provide customer support service or customer sales service. It can be used by large chain stores to connect cash registers directly with central computers, which can aggregate information on sales volume, inventory, and demand for specific items offered for sale. Every day, hundreds of thousands of persons dial toll-free numbers to order merchandise, register complaints, ask for service, or obtain information. Call center agents who answer customers' questions about their accounts or help customers with service or warranty problems or billing inquiries can have their positions equipped with screens that display the telephone number and name of the customer. The agent can access the database and have account information, history of previous calls, and other information displayed on the screen so that the caller can be answered efficiently. When Ms. Smith calls to complain about the performance of her new dishwasher, the agent can correlate her account number with the telephone number and name displayed and access the computer file of her account. She can then say, "Good morning, Ms. Smith," when she answers the call. When Ms. Smith complains about the dishwasher, she can view the account record on her monitor, see that the washer is under warranty and arrange for a service call.

CTI is also used to provide interactive voice response (IVR), which allows customers to conduct business transactions by using touch-tone telephones to respond to voice questions or directions. Examples are automatic call distribution (ACD) systems and systems that will provide banking and credit card account balance and transaction information in response to telephone keypad commands. Callers may open accounts, verify account balances, determine the date that checks cleared and the amount for which they were written, transfer funds between accounts, pay bills, and trade mutual funds. The state of Massachusetts' "Telefile" program allows taxpayers to file tax returns by telephone using IVR. Over 172,000 tax returns were filed by phone in 1995.

Newer systems provide a voice user interface (VUI), which can recognize speech and will respond to voice commands. Instead of pressing numbers on a telephone keypad, callers can respond with a word or phrase. For instance, one system provides current trading prices of corporate stocks on the New York, American, Vancouver, Calgary, and Nasdaq stock exchanges. Callers are asked to state the name of the stock exchange, then the name of the stock. The system has been programmed to recognize common variations of a stock's name including its initials. The stock prices in the system are updated continuously, and bid and asked prices of the most recent trades are quoted.

VUI is now being used in place of telephone keypads for interactive voice response in call centers. The caller can now ask for customer service instead of listening to a menu for several minutes and pressing number 5 after being told that that is how to reach customer service. In the case of travel reservations, responding to computerized voice prompts, a caller can ask for a list of flights between two points, on a specified day, within a desired time span without working through a maze of menus and keypad responses. The systems can also be directed by voice commands to reserve a rental car and make a hotel reservation. You can call a travel agency equipped with a VUI, and when the system answers tell it that you want a business class seat on a flight from San Francisco to Dallas, leaving San Francisco after 5:00 pm on November 1. You will be able to select by voice command from a list of flights presented to you. The system will ask you if you want a return flight. You can respond verbally to this and other prompts about car rental and hotel reservations. You will then be handed off to an agent who can issue tickets.

Telecommuting

In 1998, 11 million persons worked at home, and this number is expected to double by the year 2002. These workers include home-based businesses and telecommuters.

Telecommuting is one business application of the Superhighway that is being promoted as a means for reducing the number of vehicles on the streets and highways, easing traffic problems, reducing the wear and tear on streets and on automobiles, reducing consumption of fuels, and in cities with air pollution problems, reducing the amount of pollutants released into the atmosphere. Eliminating the commute to a distant workplace, or reducing it to once or twice a week, can add many productive hours a week to a worker's output, or improve the worker's quality of life by allowing the worker to have more flexible work hours and a few more leisure hours while putting in a fully productive work week. Added benefits are the elimination of the cost of driving to work each day, which can amount to several hundred dollars per month, and the cost of providing office space at the central location. The cost of office space has been estimated to be around $4,500 per person per year, which includes heat, lighting, and maintenance.

Telecommuting can usually be accomplished through a dial-up connection between the remote worker's location and the office. The POTS connection will allow facsimile transmission and reception, access to the company computer and database through a high-speed modem, and slow scan compressed videoconferencing. If the remote location is within the local calling area or the service area of the local service provider, costs will be minimal. If greater bandwidth is required for better definition videoconferencing or higher speed data transfer, high-speed digital lines or packet switched data service may be obtained from the service providers. Cable TV companies, through cable modems, can provide high-speed access at data rates up to 40 Mbps, for a monthly charge between $30 and $40.

The Virtual Office

The "Road Warriors" of today, those who must leave the office to travel throughout the United States and the world on business, no longer must leave their office behind them, but can keep in touch even when in automobiles, airplanes, and other out-of-the-office or out-of-town locations. Cellular radio, plus portable computers with modems and facsimile machines that can be plugged into modular jacks, can provide voice, data, and image transmission to and from the home office and around the world. Telephones on airplanes and in hotel rooms are equipped with modular jacks for connecting modems to the transmission facility. Some airplanes also provide a power pack to power laptop computers.

The Internet can be accessed, e-mail and voice mail can be received and answered, memos and reports can be forwarded from the field, and information can be obtained from the company database.

For instance, John Salesrep, on his way to Memphis, calls his home office from the airplane and accesses his voice mail. When he does, he finds that a good customer at his destination needs some equipment to extend the assembly line at his manufacturing facility. John calls the customer, obtains specific requirements for the equipment, and consults the catalog information stored in his laptop computer to select the products that will meet the needs of his customer. Later, after checking into his hotel, John plugs the modem for his portable computer into the modular jack provided by the hotel, calls his office, and accesses the mainframe computer to determine the availability of the equipment. He prepares a proposal and price quotation on his laptop, calls the sales department, and transfers the proposal file to a computer in the sales office. An administrative assistant prints the proposal and price quote, assembles it with specification sheets and brochures, and submits it to the sales manager for approval. The package is sent to John at his hotel by overnight courier and John makes an appointment to present the proposal to his customer the next morning.

Video Applications

New products can be introduced and promoted to sales personnel and to customers throughout the marketplace by setting up videocasts to field offices, thus eliminating the travel time and travel and living costs required for the people of interest to come to the central office location. Interactive two-way videoconferences can be used to train sales personnel on the features and benefits of the products or to teach service personnel how to service and maintain them.

Distance Learning

Distance learning networks use the Superhighway to help solve problems such as teacher shortages and low enrollments at distant schools. These networks allow information to be shared between institutions. Some networks link the schools in only one school district; others serve all of the learning institutions throughout an entire state. One network in Hawaii is planned to interconnect 360 schools on six islands.

The networks allow students to have access to the information resources of all of the schools in the network and each school to benefit from the teaching assets of the other schools. When enrollment for a requested course is too low to justify an instructor at each of several schools in a school system, classes can be taught by one instructor at a central location and be broadcast over the Superhighway to all of the remote classrooms. By equipping the remote classrooms with videoconferencing equipment and providing a switched video network, students at a distant location can interact with the instructor and the students at other locations. When experts in different fields make presentations at one institution, the program can be broadcast to other schools on the network. Teaching resources can be expanded to include scientists, business executives, government officials, and healthcare specialists. Educational programs that can be supported by urban areas can be extended to students in rural areas.

Distance learning networks facilitate the training of teachers. They allow teachers to be observed in action by a group of peers or professors who can discuss teaching techniques without disrupting the class and can join in interactive discussions with the group after the class.

Wired Campuses

Colleges and universities are installing campus-wide broadband networks that will allow students to have real-time access to the institutions' information resources, including the library; offer two-way communications for interactive teaching of language and other courses; and provide access to the Internet. The campus networks provide the students with multimedia communications with other students and faculty throughout the campus and, where statewide educational networks have been built, with the information resources of dozens of other colleges and universities. Many of the systems support shared screen computer applications and videoconferencing.

The Virtual College

New York University (NYU) has implemented a nationwide program, which they call *The Virtual College,* that links faculty based in California, New Hampshire, New York, and Washington, D.C. with students throughout the United States. Instead of using physical classrooms, students attend virtual classes, using PCs or laptop computers to access, over toll-free lines, NYU servers with stored course material. Students can post questions or comments on bulletin boards for discussion groups and can send them as e-mail to faculty and other students. The network utilizes 14.4-Kbps or 28.8-Kbps modems over POTS telephone lines (the narrowband lanes of the Information Superhighway), so that the students can access the virtual classroom even when they are traveling.

Control of Electrical Power Transmission

The huge grid of interlocked transmission and distribution systems of the power-producing utilities in the United States is controlled over segments of the Information Superhighway. Some of these segments are part of the private networks of the power utilities and some are networks of communications service providers. Computers monitor information from sensors in the grid to determine that the power systems are in synchronism and maintaining the correct voltage levels. If one section of the grid becomes overloaded because of a fault or loss of a generator so that its voltage levels begin to drop or it begins to slip out of synchronism with the rest of the grid, the computers act to warn the other sections and to disconnect the section from the grid. If the computers do not act in time or if communications between computers are lost, the entire grid can collapse, resulting in one of those huge blackouts wherein users in a region lose power.

The private networks of the utilities are not available for general public use, but they are still part of the Superhighway, and all of the users of the utilities' services benefit from their existence.

Finance

The world of finance, which includes securities investing, currency exchange, banking, commodities markets, and metals markets, could not function as it does today without the Superhighway.

Electronic Banking

Electronic banking takes advantage of the Superhighway to provide interactive services for its customers and to transfer the enormous amounts of data necessary for it to conduct its business. Direct deposits of salaries by employers and by government agencies for Social Security and other payments are made possible by electronic funds transfer orders transmitted over the Superhighway.

CTI (discussed earlier) allows customers to open accounts, transfer funds, pay bills, and monitor check clearances, credit for deposits, automatic teller machine (ATM) withdrawals, and bank balances through IVR systems. Telecommunications also makes it possible to withdraw funds from a bank account through the use of ATMs in cities around the globe.

Looking to the future, additional banking services, loan processing for example, are expected to be available through the facilities of the Superhighway.

Investment Transactions

The hundreds of millions of transactions that are executed in the stock and bond markets of the world would be impossible without the nearly instantaneous communications between the investors and the brokers and marketers. An order to sell a corporate stock can be placed, the sale consummated, and the selling price reported within a matter of minutes.

Telemedicine

Initially, use of the Superhighway in medicine was limited to telephone conversations between doctors conferring on diagnosis and treatment for a patient. Today, however, medicine takes advantage of the availability of videoconferencing, teleradiology systems that can forward x-rays, and image transmission systems that can transport high-resolution video color images, photographs, ultrasound scans, electro-cardiograms, and electroencephalograms. Copies of health records, lab reports, and analyses can be forwarded by facsimile, or if digitized in a computer, directly by electronic data interchange (EDI). Some of the medical applications of the Superhighway are listed in Table 3-1.

Table 3-1   Telemedicine Applications of the Information Superhighway.

• Telediagnosis	Patient record access
• Teleradiology	Laboratory report transmission
• Telepathology	Remote surgery supervision
• Tele-endoscopy	Mobile patient monitoring
• Remote EEG/ECG	Medical information service

Through the transmission of images and records, specialists in other cities can be consulted for second opinions on the diagnosis, prognosis, and recommended treatment. Through live, two-way videoconferencing, doctors can confer with patients in rural areas, on remote military bases, and in prisons. Attendants with the patients can install sensors on the patient so that the doctor can hear heartbeats and watch electrocardiogram scans. Instructions to surgeons performing operations at distant sites are given by specialists watching the procedures on high resolution video.

The Superhighway may be used to access databases where medical information is available. This will allow medical practitioners to keep up with new developments and to have at their fingertips the latest information on any of their patients' medical problems.

In New Hampshire, one of the largest medical centers in New England has installed a network that links it with video and imaging facilities at four community hospitals and three clinic sites throughout New Hampshire and Massachusetts. The network operates over ISDN lines provided by the RBOC. Prior to installation of the network, doctors and hospital administrators were spending a large amount of their time shuttling between the sites of hospitals and clinics.

The use of the Superhighway for telemedicine is growing, but the growth is inhibited to some extent by the cost of equipping remote sites and because many health insurance plans will not reimburse the patient for the costs.

Transportation

Today's transportation systems could not function without the telecommunications networks that allow the tracking and control of air traffic, the identification and location reporting of railroad cars, the dispatching and centralized control of trains, the monitoring of automobile traffic, and the control and adjustment of traffic lights and signals to smooth and speed the flow of automobiles.

We have become accustomed to the "help radios" that line the highways. These radios enable drivers to call for service in the event of a breakdown and to report accidents so that service or towing vehicles and, in the event of an injury, ambulances and medical help can be dispatched.

Intelligent Transportation Systems

The FCC has allocated radio spectrum and has issued rules for the Location and Monitoring Service (LMS), the General Wireless Communications Systems (GWCS), and vehicle collision-avoidance radar systems. These services are being developed as part of intelligent transportation systems (ITS). The purpose of these services is to track the location of vehicles, provide instructions when required, and engage in two-way communications regarding their status when problems occur.

Today, motorists and commuters can determine their location electronically and receive directions for traveling to a selected destination. They can also speed through toll booths without stopping to pay the toll. Technology is being deployed that will allow vehicles equipped with a special transponder that can transmit an account or credit card code to pass through bridge and highway toll plazas without stopping. One such system is operating at the Queens-midtown tunnel in New York.

Highway 407, near Toronto in Canada, is a 69-km long, multilane express toll route. The highway is equipped with Electronic Toll and Traffic Management (ETTM), which permits cashless toll collection. Regular commuters have a transponder mounted on their windshield, which communicates with vehicle sensors mounted at each entry and exit ramp of the highway. In the future, vehicles may also be able to travel the highway without a transponder. Digital cameras at each entry and exit ramp photographs the rear license plates and record the location and time of day. A toll transaction processor matches the entry and exit data and analyzes it for billling purposes.

Within the not-too-distant-future drivers will be able to place an automobile in a fully automated mode. Computers, using signals from multiple sensors, will control the steering, driving, and braking systems to maintain lane position control, speed, and following distance, and will avoid potential collisions with the aid of collision-avoidance laser radar systems.

In a 1997 demonstration sponsored by the National Automated Highway System Consortium (NAHSC) in San Diego, a string of computer-controlled automobiles drove down a 7.5 mile stretch of automated highway at a speed of 65 miles per hour with a separation of 18m between cars. In a demonstration of collision-avoidance laser radar, an automatically driven semitrailer bore down on a stopped vehicle at 65 miles per hour. The radar system detected the presence of the vehicle and brought the truck to a stop behind it.

Conclusion

We have examined the services provided as well as many of the applications and uses of the Information Superhighway today. We can't begin to visualize the services and applications that will probably be common 10 years from now.

Visible on the horizon of systems and device development, in addition to the developments already mentioned, are personal satellite telephones weighing 1 pound or less, personal lifetime telephone numbers that will be valid anyplace on the globe, radio receivers for Global Positioning Satellite information that will show the users their exact location, networks of low Earth orbit satellites that will allow mobile telephones to be reached any place on Earth, and e-mail desk telephones that will have internal storage of messages. Many other services that are beyond our imaginations today will be developed.

Bibliography

Akselsen, Sigmund, Eidsvik, Arne Ketil, and Folkow, Trine, "Telemedicine and the ISDN," *IEEE Communications,* January 1993.

Aranda, R. Rembert and Vigilante, Richard, "Re-engineering Management Training with Networked Multimedia," *Telecommunications,* July 1995.

Freeman, David J., "Bandwidth Hunger Driving High-speed Datanets," *Communications Engineering and Design,* August 1996.

ITU-T Recommendation I.121, "Broadband Aspects of ISDN," 1990.

Krol, Ed, *The Whole Internet User's Guide and Catalog,* 2nd Ed., Oral & Associates, Inc., Sebastopol, CA: 1994.

Lane, Elizabeth S., and Summerhill, Craig, *An Internet Primer for Information Professionals,* Westport, CT: Meckler Publishing, 1992.

4 Lightwaves Widen the Information Superhighway

640K is enough for anyone.

— Bill Gates, Microsoft chair, 1981.

The Evolution of Lightwave Systems

A typical lightwave system, shown in Figure 4-1, consists of a light emitting diode (LED), or diode laser light source, to convert electrical signals into optical signals; fiber optic cable for the transmission medium; optical connectors and splices to connect the segments of the fiber; and a detector to convert the

Figure 4-1   Typical Lightwave System.

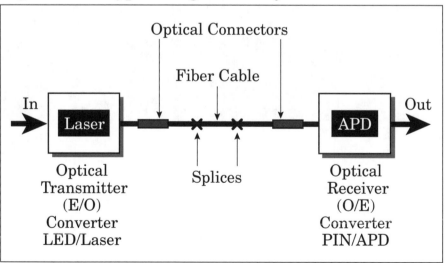

optical signals back into electrical signals. The maximum fiber lengths available commercially are about 12 km. Fiber segments are spliced together to provide longer spans when required. Connectors are used where access to the signal is required and to connect the fibers to the terminating equipment.

The signal characteristics of a lightwave system are dramatically different from those of a conventional transmission system. When dealing with ultrahigh-frequency electromagnetic waves, such as light, it is common to use units of wavelength rather that frequency (cycles per second or Hertz), as is the case with conventional systems. The frequencies used for lightwave systems extend from approximately 10^{14} to 10^{15} Hz. The theoretical bandwidth of a lightwave system is a staggering 100,000 GHz! Most fiber systems operating today have a wavelength of 1,310 or 1,550 nanometers (nm). This wavelength is beyond the sensitivity of the human eye (which cuts off at about 770 nm) in the near-infrared portion of the spectrum shown in Figure 4-2.

Figure 4-2   Optical Spectrum.

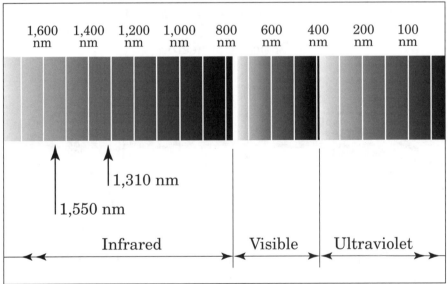

The first operational systems back in 1981 operated at a wavelength of 850 nm with a bit rate of 45 Mbps, providing a maximum capacity of 672 voice circuits over two fibers.

A simplified diagram of a typical 565-Mbps system is shown in Figure 4-3. There, 1.544-Mbps (DS-1) signals from a channel bank (24 voice circuits) are combined with 27 other DS-1 bit streams into a 45-Mbps (DS-3) bit stream by an M13 multiplexer. The resulting DS-3 stream, along with 11 other DS-3 streams, are then converted into optical pulses by the laser diode operating at a wavelength of 1,550 nm. After transmission through the fiber cable, the optical pulses are detected by an avalanche photodiode and converted back to electrical signals, regenerated and then reconverted into a standard DS-3 signal for input to the M13 multiplexers and the associated channel banks.

Figure 4-3   565-Mbps Lightwave Terminal.

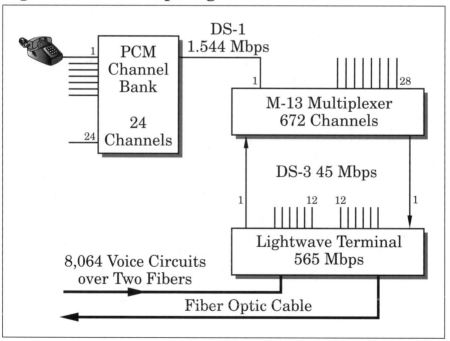

These fourth-generation fiber optic systems operate at 1,550 nm with a bit rate of 565 Mbps and provide 8,064 voice circuits per fiber pair. The systems now installed are designed to be upgraded to 2.5 Gbps. These systems could then carry four times as many calls simply by installing new electronics in terminals and repeater stations along the routes. These systems have repeaters spaced up to 100 km apart, and a bit error rate of five errors per 10 billion bits. In other words, if you transmitted the text of four 30-volume sets of the *Encyclopedia Britannica* over a distance of 100 km in 1 second, the only error would be that one letter might be capitalized instead of lowercase!

As the demand for broadband services increases, such as videoconferencing, imaging, and pay-per-view television, the telecommunication companies are upgrading their networks to deliver services at increasingly higher speeds. Today many lightwave systems are capable of supporting speeds up to 240 Gbps. This is enough capacity to carry 3 million voice circuits simultaneously over a single fiber pair.

The world's leading designers and suppliers of lightwave equipment are developing the technologies to establish new fiber-span distance records. These devices include new lasers, detectors, and other electronic and optoelectronic products. One new device is a semiconductor phase modulator that minimizes chirping at high frequencies by eliminating the need to turn lasers on and off. Chirping is the small changes in wavelength that occur when a laser is turned on and off at high speeds causing transmission errors. The modulator acts as a shutter that rapidly opens and closes, producing very short light pulses with minimal frequency distortion at 10 Gbps.

All-Optical Networks

While no one knows what new services will travel over the Information Superhighway of tomorrow, new users and new applications will bring many challenges. Network capacities will be stretched to the limit in order to deliver a range of applications with different performance and bandwidth requirements.

The solution may lie with all-optical networks, which carry information on beams of light from one end of the network to the other. Unlike today's fiber networks, which typically carry single wavelengths point-to-point across the network through intermediate stages of optoelectronic conversion, all-optical networks will multiplex, amplify, and route multiple wavelengths entirely in the optical domain without the need for conversion. This is shown in Figure 4-4, the elements of which are described below.

1.   Laser devices generate light pulses that are "tuned" to provide precise wavelengths, such as 1,533.2 nm, 1,533.6 nm, and so on. Up to 80 wavelengths will be supported on a single fiber pair in the near future.

2.   Optical modulators convert an incoming electronic bit stream into an optical signal by rapidly turning the light stream on and off.

3.   A wavelength division multiplexer combines the different wavelengths onto a single fiber.

4.   Optical post amplifiers boost the power of the outgoing signal before it is sent out across the fiber. (Boosting the power increases the transmission reach, or the distance the optical signal can travel before regeneration.)

5.   A dispersion compensation unit corrects dispersion (the "spreading" of light pulses as they travel down the fiber) to prevent unwanted interaction with adjacent pulses.

Figure 4-4   All-Optical Network.

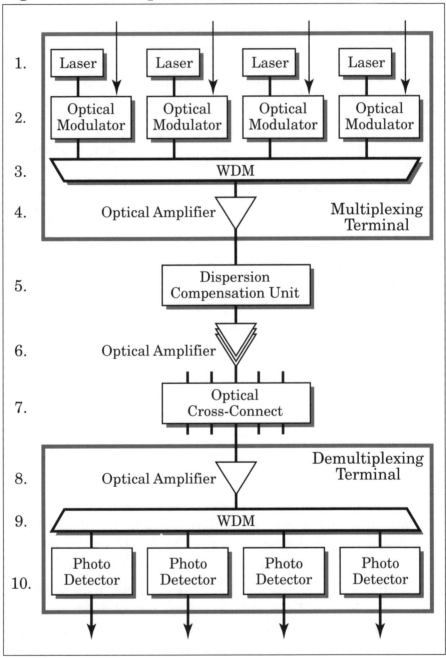

This interaction can make it difficult for the system to distinguish the pulses at the receiving end.

6. Optical line amplifiers boost signal power to compensate for losses incurred during transmission over the fiber optic link.

7. An optical cross-connect optically switches the signals to the correct destination.

8. Optical pre-amplifiers improve the effective noise performance of the optical photodetectors and boost the power of the incoming signal at the receiving end.

9. A wavelength division demultiplexer separates the multiple wavelengths carried on the incoming fiber.

10. Optical photodetectors convert the optical wavelengths into an electronic bit stream.

These all-optical networks offer the potential to deliver the huge capacities, network flexibility, and low network costs that will be required to support volume deployment of high-speed data services for Internet access and local area network (LAN) interconnection, as well as emerging high-bandwidth video distribution applications and broadband multimedia services.

Although migration to all-optical networks will be gradual, some of the elements are already being deployed. Optical amplifiers, for example, are being used to boost the power of optic signals, enabling them to travel farther without regeneration.

Optical Amplifiers

The recent development of erbium doped fiber amplifiers (EDFA) is a major milestone in lightwave communications technology. Present lightwave systems use regenerative repeaters to compensate for signal attenuation and dispersion over long transmission distances. The optical signal is converted to an electrical signal, amplified by electronic circuits, and then re-converted back to an optical signal. However, if the optical signal can be directly amplified, the repeater can be smaller and cheaper than existing regenerations.

Figure 4-5 shows the basic configuration of an erbium doped amplifier. The gain medium is fiber doped with a very small amount of a rare earth ion, erbium. The optical signal is oper-ated in the 1,550-nm window. To induce gain in the doped fiber, the optical signal is optically pumped by a laser diode. The optical pump power is combined inside the fiber core by a WDM. The pump laser operating at 980 nm offers the best absorption band for the erbium ion. The optional optical isolator is used to avoid optical feedback and laser oscillation. A gain of 30 dB

Figure 4-5   Optical Amplifier.

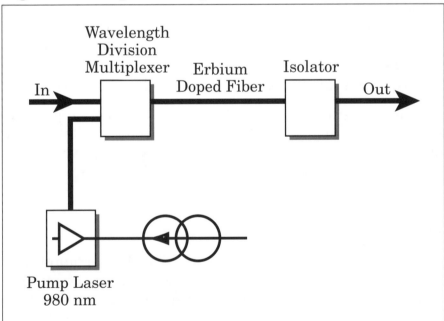

is possible. An optical amplifier can be used as a post-amplifier to increase the output power of the transmitter, as an in-line amplifier between two fiber spans, or as a pre-amplifier in front of the receiver.

Increasing the distance, or span, between regenerators allows telephone companies to extend the reach of their high-speed SONET and significantly reduce their regenerator equipment and site costs.

Wavelength Division Multiplexing

Recently, telephone companies have started volume deployment of wavelength division multiplexing (WDM) technology to increase the capacity of their existing fiber routes, thereby avoiding the cost and delay of adding new fiber optic cable to upgrade capacity. Although WDM has been used in a few applications for a number of years, the introduction of such technologies as bidirectional, multiwavelength optical amplifiers is now making WDM deployment economical.

WDM enables multiple wavelengths to be carried on the same fiber, and perhaps thousands of different wavelengths in the future. In WDM systems, lasers produce light of different wavelengths, which are then multiplexed onto a fiber to carry information across the network. Each wavelength occupies a different position in the optical spectrum, such as 1,533.2 nm or 1,533.6 nm. (See Figure 4-6.) Although single-wavelength systems operating at 2.5 Gbps and 10 Gbps are very common today, manufacturers are experimenting with systems capable of operating into the terabit (trillion bit) range and beyond.

Advanced components (such as lasers, detectors, and modulators) that support high-speed electronics and dense wavelength division multiplexing (DWDM) are critical to pushing the lim-

Figure 4-6   Typical WDM System.

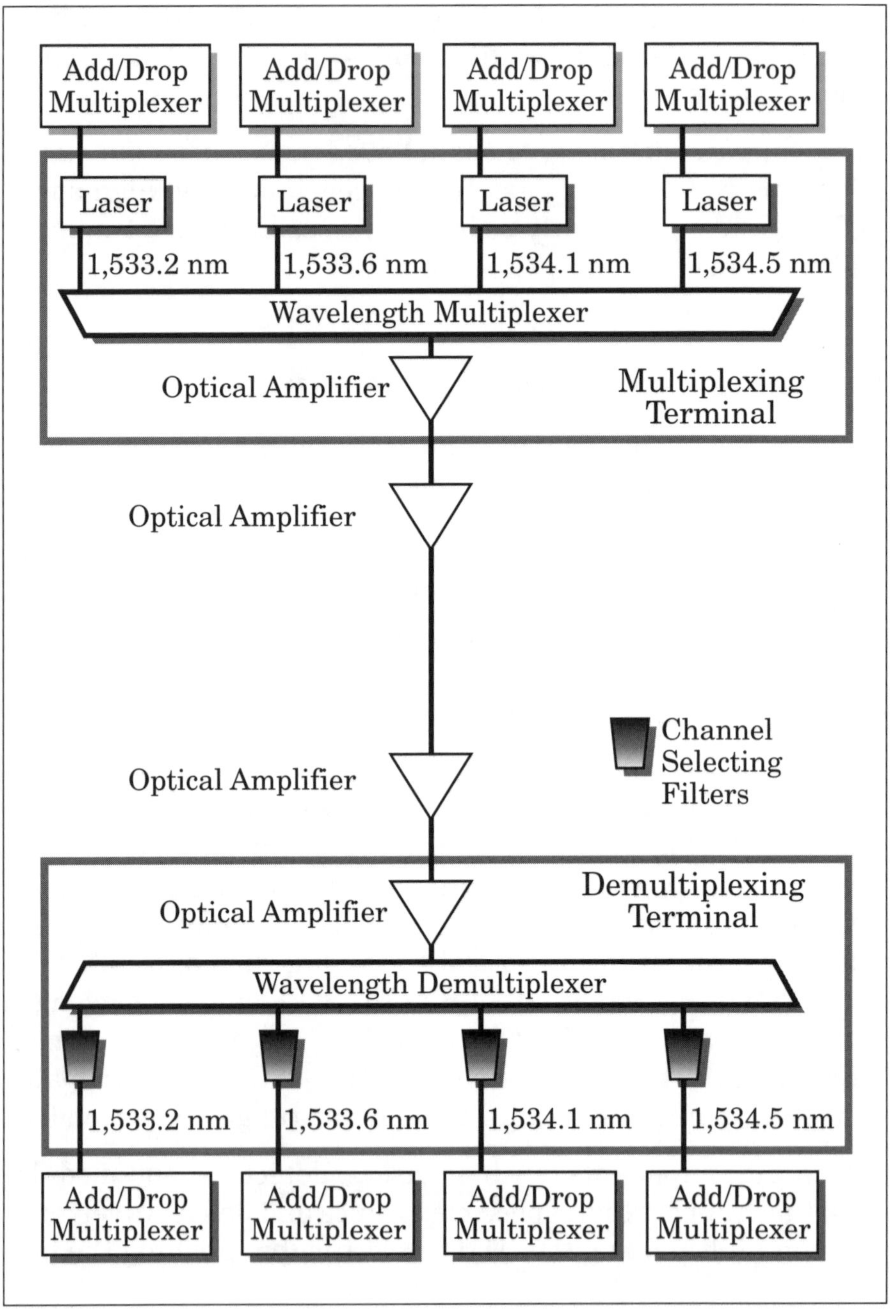

its of fiber to terabit levels, and further to 10 and possibly 20 terabits on a single fiber over the next decade. DWDM technology increases capacity by putting more wavelengths on each fiber, while high-speed electronics place more bits on each wavelength.

Nortel Networks, the world's leading provider of high-capacity optical networks, and the second largest manufacturer of optoelectronic components, can provide 320 Gbps per fiber, the highest available per-fiber capacity in the world. Based on further optical networking technology advances, Nortel Networks has committed to a terabit per fiber by the year 2000.

Meanwhile, Nortel Networks researchers are exploring several technologies that will continue to raise total fiber capacity. For example, they are investigating a technique, called soliton pulses, that could achieve in excess of 100 Gbps on a single wavelength. Soliton pulses can support higher line rates because the light does not spread as it travels down a fiber.

Optical Cross-Connects

The other main element of an all-optical network is the optical cross-connects. They will switch and route traffic from one fiber link to another in the optical domain. Today's long-haul or metropolitan area networks are terminated and demultiplexed at intermediate nodes for electronic cross-connection at line rates of 50 Mbps or lower, then remultiplexed and retransmitted onto the fiber at rates of up to 2.5 Gbps (Figure 4-7). By contrast, optical cross-connects will route traffic without optoelectronic conversion as signals are transferred from one fiber to another. In the fiber span shown in Figure 4-8, all optoelectronic conversions and low-speed ports have been eliminated. Telephone companies can also save equipment and costs

Figure 4-7   **Conventional Fiber Optic Network.**

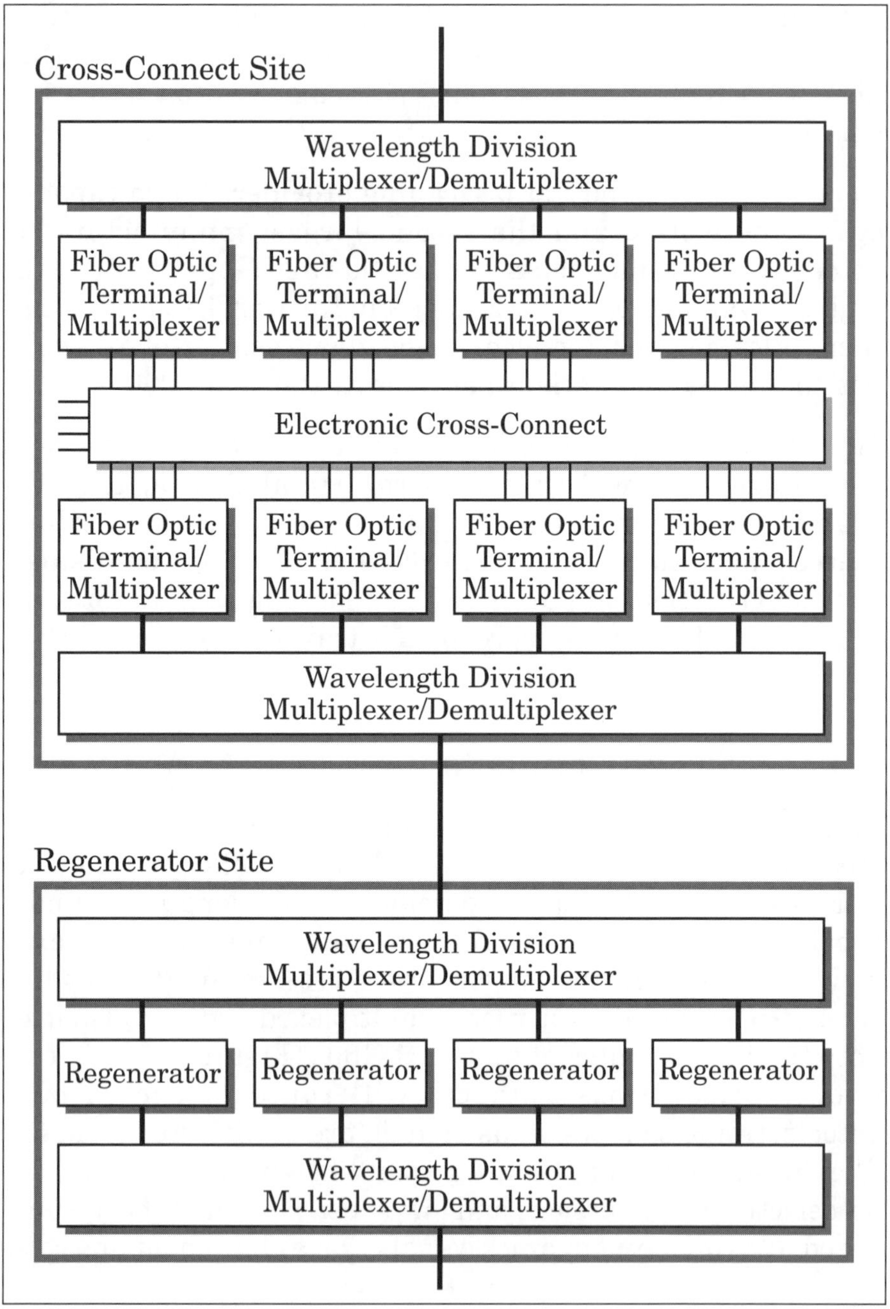

Cross-Connect Site

Wavelength Division Multiplexer/Demultiplexer

Fiber Optic Terminal/ Multiplexer	Fiber Optic Terminal/ Multiplexer	Fiber Optic Terminal/ Multiplexer	Fiber Optic Terminal/ Multiplexer

Electronic Cross-Connect

Fiber Optic Terminal/ Multiplexer	Fiber Optic Terminal/ Multiplexer	Fiber Optic Terminal/ Multiplexer	Fiber Optic Terminal/ Multiplexer

Wavelength Division Multiplexer/Demultiplexer

Regenerator Site

Wavelength Division Multiplexer/Demultiplexer

Regenerator	Regenerator	Regenerator	Regenerator

Wavelength Division Multiplexer/Demultiplexer

Figure 4-8   All-Optical Network.

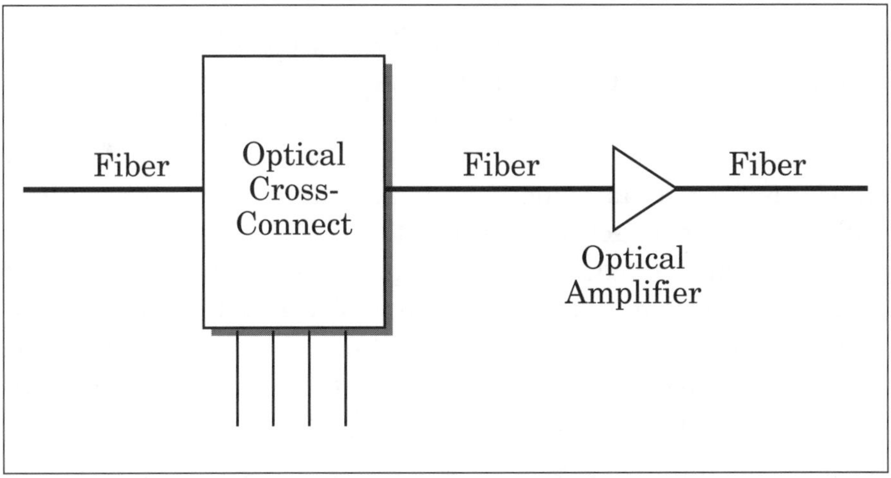

through the use of optical amplifiers. In today's conventional fiber optic networks, regenerators are used to receive, electronically reshape, retime, and retransmit optical signals to compensate for losses and distortion that build up with travel.

In all-optical networks, optical ampifiers do not reshape the optical signals but simply amplify them, allowing operators to reduce the amount of regenerator equipment and number of regenerator sites they must deploy across optical spans. Optical amplifiers, which are already being deployed in fiber optic networks, are physically smaller and more cost-effective than equivalent regenerators, especially when used with multi-wavelength applications. All-optical networks may also be able to carry and optically switch signals independently of the format (SONET, ATM, digital, analog, or video), a capability telephone companies could use to offer new services, such as the transport of cable television signals.

Conclusion

Current 10-Gbps SONET OC-192 systems can operate up to 140 km (87 miles) on a single span, and considerably farther on multiple spans where optical line amplifiers are used. These new systems will widen the Information Superhighway for such emerging services as remote medical imaging, remote business and banking transactions, Internet, and LAN access. This system, for example, could transmit 5,000 ultrasound images or 10 cardiological angiograms (videos of human blood vessels) across the United States in just 1 second! Or they could simultaneously transmit 130,000 voice calls or 21,500 channels of high-quality video conferences on a single optical fiber.

By the year 2000, Internet traffic could account for more than half the capacity on long-haul telephone networks. In 1996, it made up less than 1%. In the next few years, network capacity demands will grow enormously as more people sign onto the Net, spend more time on it, and access higher-bandwidth interactive multimedia services. This explosive mix will push long-haul fiber capacities into the terabit range and beyond by the year 2000.

Optical technologies will light up the Information Superhighway of the future to cost-effectively support volume deployment of new high-speed data services and high-bandwidth video and multimedia applications.

As the year 2000 approaches and the Information Superhighway becomes a reality, lightwave systems will be capable of carrying all the interactive services that the public will need.

Bibliography

McFarlane, John, Miceli, Jim, Morin, Philippe, Roorda, Peter, and Scott, Mike, "Highways of Light," *Nortel Networks Telesis,* No. 101, September 1996, p. 2.

Nellist, John G., *Understanding Telecommunications and Lightwave Systems,* New York, NY: IEEE Press, 1996.

Nortel Networks, *Nortel Edge,* No. 2, 1998.

5 The Undersea Information Superhighway

Who the hell wants to hear actors talk?

— Harry Warner, Warner Brothers, 1927.

In 1953, AT&T and the British Post Office agreed to build the first transatlantic telephone (TAT-1) undersea cable. The cable was turned on for service between Nova Scotia and Scotland in 1956. This copper cable carried only 36 circuits.

In 1988, the world's first transatlantic cable (TAT-8) to use fiber optic technology was installed to provide service to the United Kingdom and France. TAT-8 carried 8,000 circuits. TAT-9, installed in 1991, was capable of carrying 80,000 simultaneous telephone calls.

The TAT-12/13 Network

A new undersea cable network linking North America and Europe was turned on in 1996 and was named TAT-12/13.

Figure 5-1 shows a map of the TAT-12/13 Network. The network consists of a ring of undersea cable segments interconnecting cable stations in Green Hill, Rhode Island; Lands End, England; Penmarch, France; and Shirley, New York. The undersea cable segment between Green Hill, Rhode Island, and Lands Ends, England, is 5,913 km long. The segment has 133

Figure 5-1   The TAT-12/13 Cable Network.

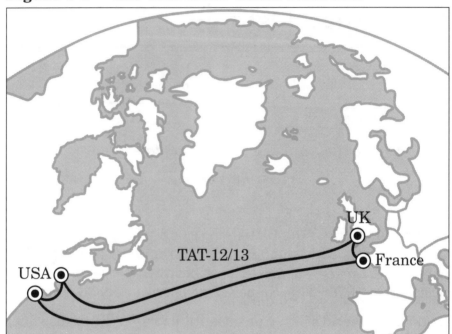

repeaters spaced every 45 km that make use of optical amplifier technology.

The undersea cable segment between Lands End, England, and Penmarch, France, is 370 km long and contains four undersea optical amplifiers spaced 74 km apart. The undersea cable segment between Shirley, New York, and Green Hill, Rhode Island, is 162 km long and contains no undersea repeaters. It makes use of high-power optical amplifiers in the terminal station equipment enabling traffic to traverse the entire segment without undersea amplification. The undersea cable segment between Penmarch, France and Shirley, New York, is 6,321 km long and contains 140 optical amplifiers spaced every 45 km. This will create a ring of undersea cable around the North Atlantic Ocean.

The TAT-12/13 Network is capable of carrying up to 10 Gbps of traffic (300,000 voice circuits) that is fully restorable within the network. By using ring switching equipment in each of the four cable stations, traffic is automatically rerouted in the opposite direction around the ring to bypass a fault in any of the undersea segments. This rerouting takes place quickly enough that service will not be interrupted, making failures transparent to the network users.

TAT-14

The latest transatlantic cable connecting the United States and Europe will be TAT-14. This cable will employ WDM technology. It will have a 640-Gbps capacity, about 80% of which will be designated for the Internet. It also will have the ability to carry 7.7 million telephone calls simultaneously. Spearheading the project are AT&T, British Telecom, Cable & Wireless, Deutsche Telekom, and France Telecom.

The cable will link Germany, the United Kingdom, Denmark, France, and the Netherlands with the United States, spanning 22,000 miles. It will be operational by the end of 2000.

Globesystem Atlantic

Globesystem Atlantic is another link in the Information Superhighway in the North Atlantic. Globesystem will form an 8,860-km broadband link connecting Europe, Canada, the United States, and points beyond. The transatlantic portion is Teleglobe Canada's 7,500-km CANTAT 3 undersea cable. With a total capacity equivalent to 60,480 voice circuits, CANTAT 3 will have six terminating points between Canada and Northern Europe, as shown in Figure 5-2.

Figure 5-2   Globesystem Atlantic Network.

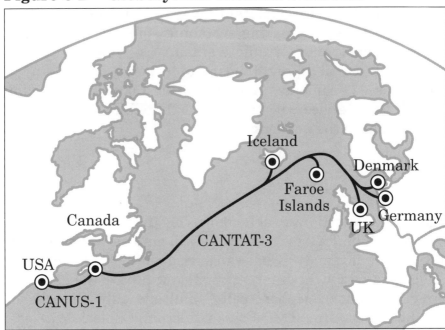

CANUS 1, an undersea cable that is 50% owned by the U.S.-based Optel Communications, Inc., will provide the 1,360-km extension from the Canadian landing point in Nova Scotia to the northeast United States. Using advanced synchronous digital hierarchy (SDH) technology, the complete Globesystem will provide users with a leading-edge platform capable of supporting interactive multimedia applications such as telemedicine, computer-aided design, and distance learning.

Pacific Ocean Undersea Cables

Undersea fiber optic cables in the Pacific Ocean include TPC-3/HAW 4, linking the continental United States with Hawaii, Guam, and Japan, TPC-4 (United States and Japan), TPC-5

(United States, Hawaii, Guam, and Japan), and NPC (United States, Alaska, and Japan).

A 7,400-km cable (Pac Rim West) connects Australia and Guam. An 8,600-km cable (Pac Rim East) connects Australia and Hawaii.

This is all part of a plan to join Malaysia, Singapore, Brunei, and the Philippines in the west; Hong Kong, South Korea, and Japan in the north; Australia in the south; and Guam and Hawaii in the center, with the United States by the latest in undersea lightwave technology.

The TPC-5 Cable Network

The latest addition to the undersea superhighway in the Pacific Ocean is the TPC-5 Network. The TPC-5 Cable Network is the first self-healing transpacific ring network. It consists of a ring of undersea cables connecting six network nodes at six cable landing sites. Two landing sites are on the U.S. mainland at Coos Bay, Oregon, and San Luis Obispo, California; two are in Japan at Ninomiya and Miyazaki; and two mid-Pacific landing sites are at Tumon Bay, Guam, and Keawaula, Hawaii. The total cable route around the ring is more than 24,000 km. Figure 5-3 illustrates the TPC-5 topology. As shown, the six landing sites terminate six undersea cable segments, G, H, I, J, T1, and T2, with respective lengths of 4,200, 6,580, 2,920, 8,600, 1,170, and 1,050 km.

The TPC-5 is owned by a consortium of 65 international telecommunications service carriers. These owners contracted with AT&T Submarine Systems, Inc., KDD Submarine Cable Systems, and Toshiba Corporation to build the network at a cost of $1.6 billion.

Figure 5-3   TPC-5 Cable Network.

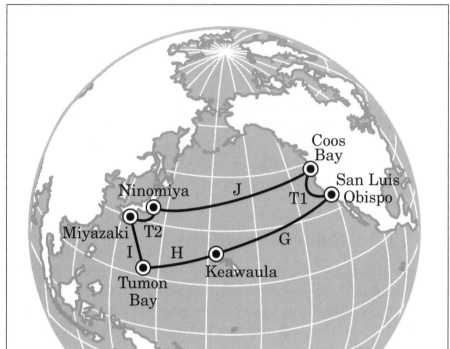

The TPC-5 brings together a unique blend of technology to achieve the transmission performance needed by network users. Among the technologies used in the TPC-5 are optical amplifiers, 5-Gbps line terminating equipment (LTE), line monitoring equipment (LME), SDH STM-16 add/drop multiplexers (ADMs) with four-fiber-ring automatic protection switching capability, and network management equipment (NME). The TPC-5 is the largest operating ring network in the world.

The Asia Pacific Cable Network

The Asia Pacific Cable Network (APCN) is a regional undersea telecommunications network. APCN interconnects Japan,

Korea, Taiwan, Hong Kong, Singapore, Indonesia, Malaysia, Thailand, and the Philippines with an 11,500-km trunk and branch undersea cable system. APCN is owned by a consortium of telecommunication carriers and was built by AT&T Submarine Systems Inc., Alcatel Submarine Networks, and KDD Submarine Cable Systems at a cost of $540 million.

Figure 5-4 shows the geographical route of APCN. The nine landing sites are connected by undersea cables and branching units. The network topology consists of 13 digital line segments that provide fiber pair transmission paths between these points. Two fiber pairs in the undersea cable terminate at 5-Gbps LTE. Each LTE transmits and receives the 5-Gbps optical line signals, composed of two bit-interleaved STM-16 (2.5-Gbps) signals. APCN will interconnect countries in a region of the world that is experiencing rapid growth in demand for telecommunication services.

The FLAG Cable System

A $1.5 billion project to create an Information Superhighway accessible to three-quarters of the world's population has been completed. The FLAG (fiber-optic link around the globe) cable system (Figure 5-5) is the longest manmade structure ever assembled, stretching 27,000 km from Porthcurno in Cornwall, England, to Miura in Japan, with landing points in Europe, the Middle East, Africa, and Asia.

The system is being built by a consortium of AT&T Submarine Systems of the United States and KDD Submarine Cable Systems of Japan. FLAG will connect 12 countries with more than 120,000 voice channels via 27,000 km of mostly undersea cable.

Physically, FLAG consists of four glass-fiber strands,

Figure 5-4   Asia Pacific Cable Network.

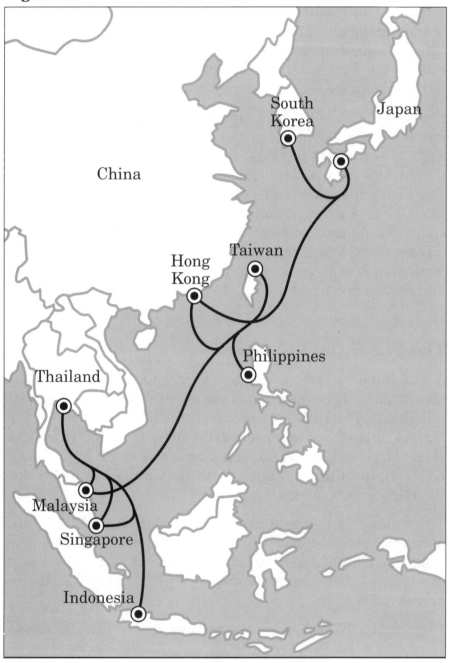

Figure 5-5   FLAG Undersea Cable System.

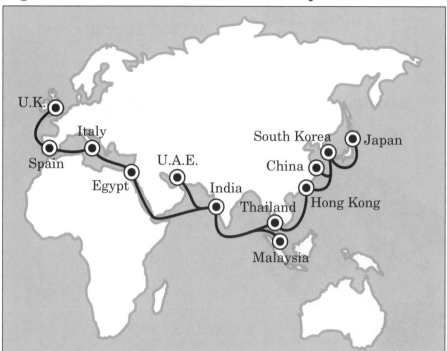

surrounded by armor to protect them against everything from fishing gear to inquisitive sharks (Figure 5-6). FLAG has two terrestrial crossings, one in Egypt and one in Thailand. Each of these land crossings will be fully duplicated on completely diverse routes so that any fault with one route will cause automatic protection switching to the other route, with a switching time of less than 50 ms.

The eastern spur of the duplicated Egyptian crossing will contain a submarine cable system between Alexandria and Port Said. This eastern spur is then continued down to Suez using an overland fiber optic cable in ducts. The full 10-Gbps capacity is carried over two fiber pairs, each operating at 5 Gbps. The western spur of the Egyptian crossing is entirely terres-

Figure 5-6   FLAG Undersea Cable.

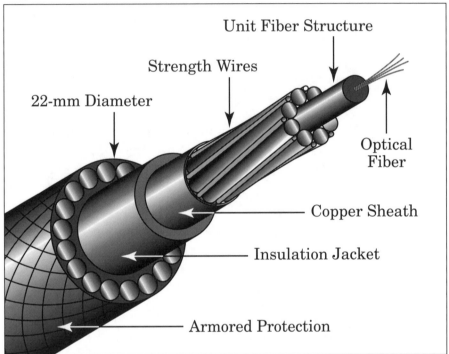

trial. This route links Alexandria and Suez through Cairo, where optical signals are regenerated. The full 10-Gbps capacity is also carried by a terrestrial fiber optic cable in protective cable ducts.

A land crossing in Thailand was necessary because of both the hazardous marine environment in the Malacca Straits and the proximity to other cables in that region. This crossing will also be done using two diverse routes. One route is 180-km long and goes from Pak Bara (Satun) on the west coast to Songkhla on the east coast. The other route is 295 km and goes through the town of Trang, where an optical amplifier repeater station will be used. Both cables will be installed in protective cable ducts where conditions are suitable.

FLAG has been funded privately by a consortium of Nynex Network Systems of the United States, Dallah-Al Baraka Group of Saudi Arabia, the Asian Investment Fund of Hong Kong, Telecom Holding Co. of Thailand, Marubeni Corp. of Japan, and Gulf Associates and GE Capital of the United States.

Some 50 telecom carriers from 45 countries, including AT&T and Sprint of the United States and KDD of Japan, have already agreed to purchase capacity on the cable.

The cable will be able to carry sophisticated traffic including medical imaging, long-distance learning, videoconferencing, multimedia, and high-definition television.

Many of the regions it will serve currently depend on satellite transmission. Fiber optic cable, however, provides increased security, speed, and accuracy of transmission as well as the capacity for advanced and two-way transmissions.

Sea-Me-We3

Another mega-project, similar to FLAG but even longer, is the $1.73 billion Sea-Me-We3 cable, approved in January 1997 by the 70 countries involved in its construction. Stretching 38,000 km, it will connect Southeast Asia, Western Europe, and the Middle East. It will employ WDM technology with eight wavelengths on two fiber pairs, giving it a capacity of 40 to 80 Gbps. It will be operational in 1999.

The Africa Optical Network

The Africa Optical Network (Africa ONE) will encircle the entire continent of Africa with an undersea fiber optic ring network (Figure 5-7). Using a combination of WDM and SDH

Figure 5-7   Africa Optical Network (Africa ONE).

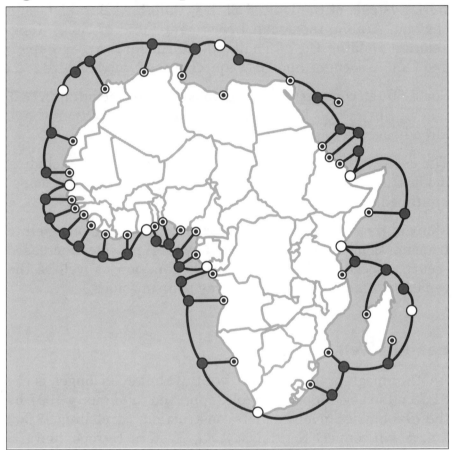

multiplex and cross-connect equipment, Africa ONE brings together a unique blend of technology to achieve network robustness.

Africa ONE is a 40,000-km trunk and branch network that is planned to be ready for service in 1999. The entire Africa ONE project integrates multiple telecommunications technologies, such as satellite, digital radio, and terrestrial fiber optic networks for providing connectivity between all African countries.

The major portions of Africa ONE will be supplied jointly by AT&T Submarine Systems and Alcatel Submarine Networks at an estimated cost of $2.6 billion.

Africa ONE is the largest undersea network ever conceived. The network will use 40,000 km of undersea fiber optic cable, encircle the second largest continent in the world, and have landings in approximately 40 different countries. The network will automatically restore traffic in the event of a failure and will interconnect African nations to each other and to the rest of the world with a world-class network using innovative WDM, optical amplification, and SDH technology. Africa ONE is a telecommunications network capable of carrying voice, video, and data services. For example, because it has adequate bandwidth to carry distance learning and telemedicine services, Africa ONE will also bring advances in education and medicine to African countries. This project, however, will do more for Africa than simply carry telecommunications services; it will stimulate economic development via the investment community. This will generate work and help to supplement basic infrastructure development. New business opportunities in telecommunications and other related fields will result.

Africa ONE will not only serve as a regional network, but as a feeder network to other networks, providing connectivity to other parts of the world. Africa ONE will play an important role in the ever-growing global undersea Information Superhighway.

Brazil's Atlantis II Undersea Cable System

In Brazil, a group of nine telecommunications companies plan to build a 12,000-km fiber optic submarine system linking South America, Africa, and Europe, designated as Atlantis II. When completed in 1999, the system will close a fiber ring in the Atlantic Ocean and provide Brazil and other South American countries with a redundant fiber link to Europe (Figure 5-8). Total project cost is expected to reach $320 million.

Vietnam, Thailand, Hong Kong Undersea Cable System

The Vietnam, Thailand, Hong Kong (V-T-H) cable link was completed in 1996. Cable terminal landing points are at Si Racha, Thailand, on the northeast shore of the Gulf of Siam, and at Deepwater Bay in Hong Kong. For the Vietnamese terminal,

Figure 5-8   Atlantis II Undersea Cable System.

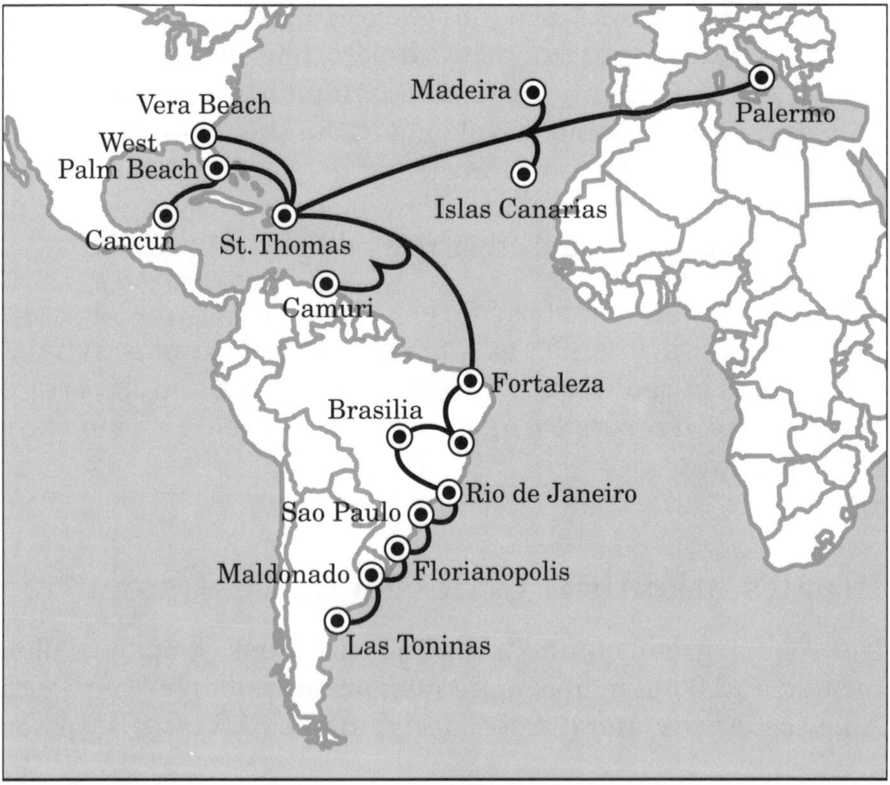

the cable landing point is in the coastal province of Vung Tau, near Ho Chi Minh City (Figure 5-9). The V-T-H cable is operated jointly by the Communications Authority of Thailand, Hong Kong Telecom, the Vietnam Posts and Telecommunications (VNPT), and the Australian company, Telstra. The undersea cable comprises two line pairs with a reported aggregate capacity of 7,000 calls each way of 565 Mbps. Fujitsu and Alcatel Submarcon installed the cable.

Figure 5-9   Vietnam, Thailand, Hong Kong Undersea Cable System.

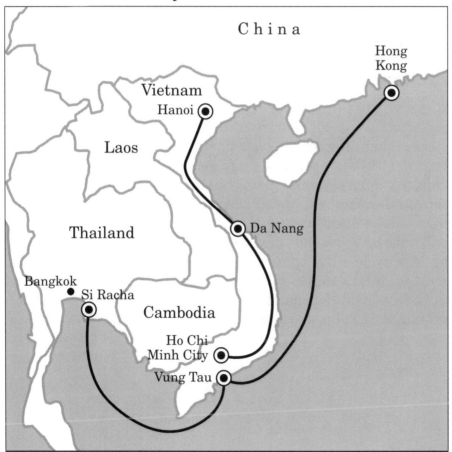

Conclusions

From 1987 to 1996, the total amount of undersea fiber optic cable installed exceeds that of copper cable installed during the 37 years from 1950 to 1987. One hundred and twenty-nine countries are connected to undersea fiber optic cable networks. Cable capacities can be upgraded in terrestrial systems by changing the repeater equipment along the routes while leaving the fiber in place. But until recently, capacity upgrades have not been possible in repeatered fiber optic undersea systems. Unrepeatered systems have long been promoted as upgradable because they do not employ underwater electronics. Increases in capacity are accomplished by upgrading terminal equipment to higher bit rates as distance and system considerations allow.

But the current generation of optical amplifiers offers repeatered systems the potential for upgradability with proper system planning at the outset, such as was done for several recent 10-Gbps systems. Two basic methods for achieving capacity upgrades are using faster operating transmission terminal equipment and WDM technology. By the year 2000, many fiber optic undersea systems using WDM technology will have 32 wavelengths on a single fiber, each carrying 5 Gbps, or 16 wavelengths operating at 10 Gbps. In the short-term, most undersea routes cabled with these systems are expected to have adequate capacity to meet requirements of the Information Superhighway during the next 5 years.

Bibliography

"Global Undersea Communication Networks," *IEEE Communications,* February 1996, pp. 24-28, 30-40, 42-48, 50-57.

Denniston, Frank J., and Runge, Peter K., "The Glass Necklace," *IEEE Spectrum,* October 1995, pp. 24-27.

Hecht, Jeff, "Planned Super-Internet Banks on Wavelength Division Multiplexing," *Laser Focus World,* May 1998, pp. 103-105.

6 The Superhighway in the Sky

There is no likelihood man can ever tap the power of the atom.

— Robert Millikan,
Nobel Prize winner in physics, 1920.

A Brief History

The first sounds to be transmitted from outer space by a man-made device were "beeps" from the Russian *Sputnik 1* launched on October 4, 1957. The United States entered the space age on January 31, 1958 with *Explorer 1*.

The United States placed the first communications satellite in orbit a year later. *Score*, a short-lived but highly successful satellite, relayed messages up to 3,000 miles and broadcast to the world a tape-recorded Christmas greeting from President Eisenhower.

The concept of a geostationary satellite was first published in October 1945 in an article entitled "Extraterrestrial Relays," by British scientist Arthur C. Clarke. Although his article seemed pure science fiction to many at that time, less than 20 years later the world's first commercial communication satellite was launched on April 6, 1965, by the Communications Satellite Corporation (Comsat) in the United States. Comsat represents the United States in the International Telecommunications Satellite Consortium (Intelsat).

This synchronous satellite, named Intelsat 1, or Early Bird, was placed in geostationary equatorial orbit over the Atlantic Ocean (Figure 6-1). From its vantage point 22,300 miles above the Atlantic, Early Bird linked North America and Europe with 240 high-quality voice circuits and made live television commercially available across the Atlantic for the first time.

Figure 6-1

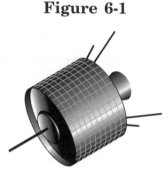

The launching of early communication satellites was an extremely costly operation using Thor-Delta and other series of rockets. The Thor-Delta launch vehicle, shown in Figure 6-2, was 116 feet long and 8 feet in diameter. However, the reus-

Figure 6-2   The Thor-Delta Launch Vehicle.

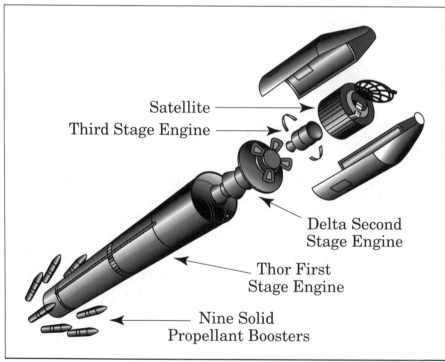

Satellite
Third Stage Engine

Delta Second
Stage Engine

Thor First
Stage Engine

Nine Solid
Propellant Boosters

able space shuttle, shown in Figure 6-3, is a manned vehicle that can place very large payloads in orbit more economically. The space shuttle is put into circular orbit at 320 km (200 miles). To launch a satellite, a system is used to spin the satellite to 60 rpm and then

Figure 6-3

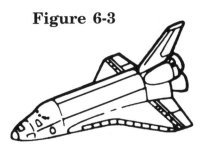

eject it by means of springs. The satellite coasts clear for 45 minutes before the solid propellant motor is ignited.

Geostationary (GEO) Satellites

To achieve orbital velocity, a satellite must be lifted above the atmosphere and propelled around the Earth at a speed that will produce a centrifugal force equal but opposite to the gravitational force at that altitude. If the speed is too fast, the satellite will fly off into space. If the speed is too slow, the satellite will be pulled back to Earth by gravity. Figure 6-4 illustrates the principle of orbital velocity. When velocity, direction, and gravitational force balance, the satellite "falls" in a circular orbit. The Earth's gravitational attraction decreases with altitude. Therefore, high-altitude satellites do not have to circle the Earth as fast as low-altitude satellites.

At an altitude of 35,880 km (22,300 miles), a satellite obtains a stable orbital speed of 6,870 miles per hour. At this speed, it orbits the Earth in exactly 24 hours. This is a geosynchronous satellite because its orbit is synchronized to the Earth's rotation. A synchronized satellite, placed in orbit directly above the equator on an eastward heading, appears to be stationary in the sky, so its orbit is referred to as a geostationary (GEO) orbit.

Figure 6-4   Orbital Velocity.

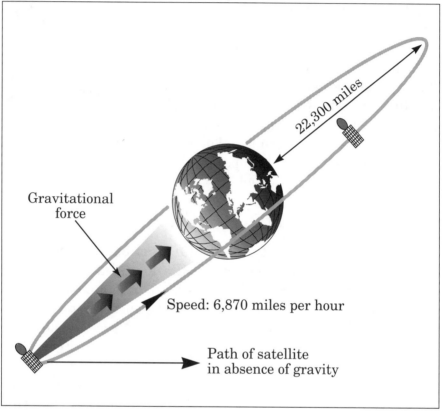

Before the late 1970s, nearly all communications satellites used the 6/4-GHz band. The 4-GHz band was used for the downlink, and the 6-GHz band was used for the uplink. Used together they are called the C-band. Because the frequencies of the band coincide with those used for terrestrial microwave systems, their applications were limited, due to radio interference. However, the more powerful "second-generation" satellites operating in the 14/12-GHz band are capable of utilizing smaller and less expensive antennas. This band is referred to as the Ku band.

Parking Slots

Satellite systems have to make use of limited spectrum allocations. In addition, satellites must have sufficient spatial separation to avoid interfering with each other, so there is a limit to the number of satellites that can be parked in a given section of geosynchronous orbit. These satellites operate either in the C band (6/4 GHz) or Ku band (14/12 GHz).

Ku-band Earth station antennas typically are 1.8 to 2.4m in diameter. They are small enough to be located on the roofs of buildings without interference from existing microwave radio systems in the same area (Figure 6-5). The existence of the very small aperture Earth terminal (VSAT) has made it possible to bring satellite communication directly to the end user's location.

Figure 6-5   Typical Earth Station Antenna.

Under regulations established by the U.S. FCC in 1974, U.S. communications satellites had to be positioned 4 degrees apart (about 2,900 km or 1,800 miles). As a result there were only 13 possible locations from which signals could be transmitted to the United States. The agency increased the total to 32 locations in 1977 by allowing the 14/12-GHz satellites to be positioned 3 degrees apart. By requiring tighter control of transmitter power and antenna patterns, the FCC has since reduced the spacing to 2 degrees, thereby increasing the number of allowable parking slots. Presently, there are almost 100 geosynchronous satellites worldwide. Covering North America, there are 14 C-band and 11 Ku-band satellites, and an additional 9 satellites with both C-band and Ku-band capacity.

Transponders

The electronic circuitry on the satellite is called a transponder. It receives the signal transmitted from the Earth station, amplifies the signal, changes the frequency, and retransmits the signal back to Earth. Each radio channel has its own transponder, so a number of transponders are on board the satellite to cover the allocated frequency band. Current communications satellites typically have 40 transponders, each with 36 MHz of usable bandwidth. A single transponder can carry one color television signal, 1,200 voice circuits, or digital data at a rate of 50 Mbps.

On-board switching may be required on satellites in the future to assign bandwidth for a complex mix of multimedia services, and to make the best use of the beams linking the satellite to the ground. By contrast, today's conventional satellites do not contain switches, so that they can conserve scarce space and satisfy weight restrictions. Instead, they are simple "bent-pipe" vehicles that receive signals from the Earth and reflect them back to ground-based switching platforms. Dynamic bandwidth management will be critical to handling the future traffic patterns and grade-of-service characteristics (such as delay-sensitive or bursty) of multimedia services. Using dynamic allocation techniques, satellites could rapidly re-assign more bandwidth to a high-demand service, for example, a large Internet download, and then shift bandwidth to other users as new demand arises.

The Footprint

The size and shape of the satellite's radio beam on the surface of the Earth is called the footprint. The actual antenna design and the transmitting power are important factors in determining the size of the footprint. The satellite can send narrowly focused spot beams to specific Earth stations with a very small footprint. However, if the satellite is broadcasting television signals directly into homes, the footprint would be very large (Figure 6-6).

Figure 6-6   Satellite Footprint.

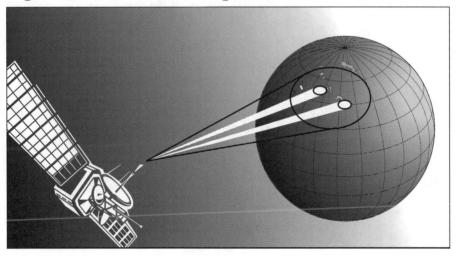

Time Delay

Signal transmission time, or propagation delay, and the associated echo become significant with satellite communications. The minimum distance between any two points via a satellite is 22,300 × 2, or 44,600 statute miles (72,000 km). Consequently, a Los Angeles to Chicago circuit via satellite will have a round-trip delay of 540 ms compared with the longest coast-to-coast terrestrial connection of about 50 ms (Figure 6-7). Echo cancellers can control the echo, but nothing can be done about the delay.

With the widespread introduction of domestic satellite systems into the network, there is a possibility of satellite links being

Figure 6-7   Transmission Time Delay.

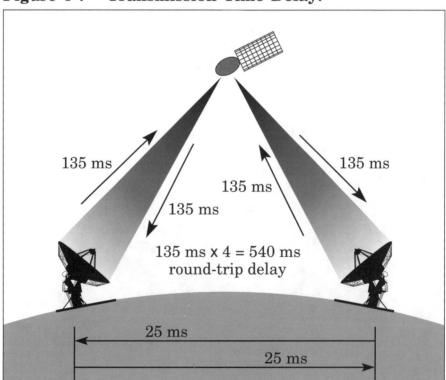

135 ms

135 ms

135 ms

135 ms

135 ms × 4 = 540 ms
round-trip delay

25 ms

25 ms

116

connected in tandem. Such double-hop satellite circuits should be avoided because the 1,080-ms delay (2 × 540 ms) is unacceptable to most people. While the delay problem is a major disadvantage for telephone circuits, satellites are ideal for one-way television transmission on transoceanic links and for domestic TV distribution.

MSAT®

One of the world's first mobile satellite systems to deliver mobile communications services directly from satellites to hand-held subscriber or vehicular access units was deployed in 1994. The system, called MSAT®, operates in the 1- to 2-GHz L-band frequency and utilizes two satellites. One satellite is owned by the American-Mobile Satellite Corporation and the other by TeleSat Mobile Inc. in Canada.

This system enables a trucking company to send information to a driver on the Trans-Canada highway in Ontario and another on the Los Angeles freeway at the same time. This has created a multi-nationwide system of mobile communications for American, Canadian, and Mexican users.

MEO and LEO Satellites

Mobile telecommunications operators favor the use of satellites operating in low Earth orbit (LEO) and middle Earth orbit (MEO). The LEO satellites operate at altitudes of about 1,000 km, and MEO satellites operate at altitudes of about 10,000 km. From these locations, satellites can concentrate a satellite signal and avoid the transmission delays and echoes that affect GEO satellites. Users can access LEO satellites via hand-held, low-powered terminals that will approximate the size of cellular radio telephones. However, the stronger signal achieved by closer proximity to Earth reduces the scope of geographical coverage and necessitates a larger number of satellites to operate in a global constellation. Nongeostationary orbital locations mean that satellites will speed across the horizon rather than appear stationary. These moving targets require complex tracking and coordination.

New mobile satellite service ventures take advantage of the still widespread geographical coverage from LEO and MEO satellites. (Refer to Chapter 9 for details of the Global PCS Network.) Operators of these networks will concentrate on providing voice and slow-speed data services anytime and anywhere, services not yet feasible with wire line facilities given the expense of serving remote areas and locales with rugged terrain.

A Failure to Communicate

Canada had the first domestic geostationary satellite system. Beginning in November 1972, a series of satellites, designated Anik A, B, C, D, and E, were launched to carry telephone traffic and video services across Canada. The system is owned by Telesat Canada, which includes a consortium of Canada's nine major telephone companies and Spar Aerospace Ltd. From 1972 to 1991, Telesat experienced only two technical failures, which were fixed within a few months.

However, on March 26, 1996, Anik E1 experienced a failure that caused the satellite to lose about two-thirds of its channel capacity. Telesat has plans to launch the next generation of satellites in two of Canada's eight spare slots. These satellites could beam down signals over a geographical area that runs from Alaska to Mexico and from Hawaii to Puerto Rico. But this serious failure resulted in a shortage of transponder space and delayed the introduction of direct-to-home television service in Canada.

Satellite failures are rare but they do occasionally happen. In May 1998, the Galaxy IV satellite suffered a malfunction that resulted in millions of pager users in United States losing contact with the satellite that relays their messages. The malfunction could not be fixed because it involved the failure of the satellite's on-board control system and backup switch.

The Death Star

One technology that has proven to be very popular is the direct-to-home (DTH) satellite system. DTH has been available in the United States since 1994. DTH or "Death Star," as it has been dubbed, is an extraordinary technology. A satellite produces signals strong enough to be received by a satellite dish antenna only 18 to 36 inches in diameter. The signals are decoded by a box the size of a CD player, which plugs directly into your TV (Figure 6-8). It can deliver more than 400 channels of audio and video, all of them digitized. Not only are there more channels possible than on cable, but the signals are technologically superior, providing laser disc-quality pictures and CD-quality audio.

Satellite Operators

There are currently five U.S. operators of direct broadcast satellite (DBS) services. The largest operator is DirecTV. This is a partnership of Hughes Electronics and RCA. DirecTV, based in El Segundo, California, has its uplink center in Castle Rock, Colorado. The other four operators are Primestar, AlphaStar,

Figure 6-8   Direct-to-Home (DTH) Satellite System.

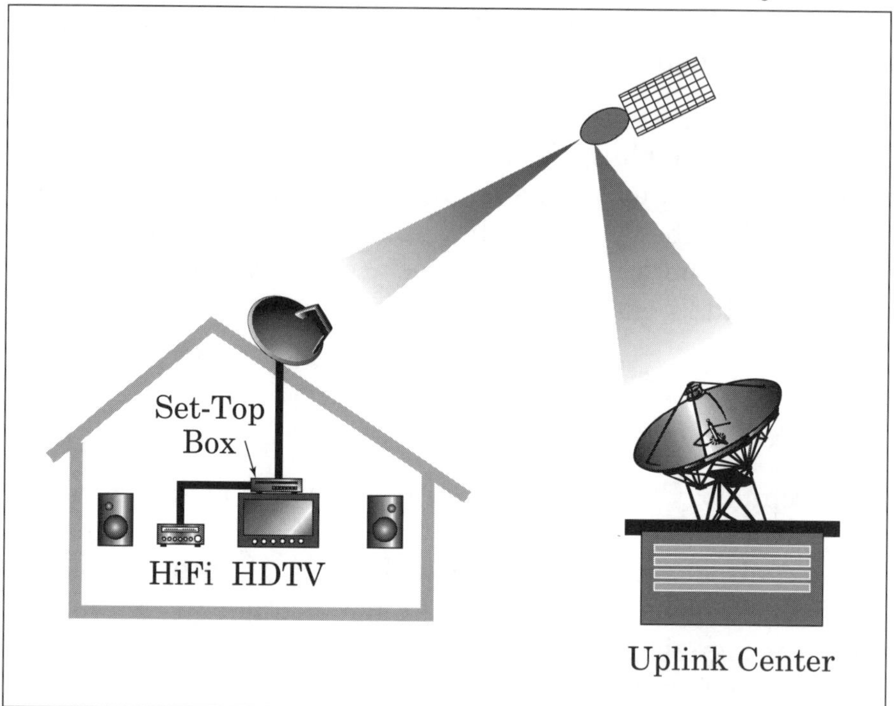

EchoStar, and USSB. Primestar, which is owned by the five leading cable companies and General Electric, was established to hamper incursions by DirecTV and others into the mainly rural areas of United States that are too remote for economical service by underground cables. DirecTV, USSB, and EchoStar use 18-inch receiver dishes, while Primestar uses a 36-inch dish and AlphaStar a 30-inch dish. They provide access to scores of TV channels, hundred of movies, sporting events, and music channels for $20 to $60 a month.

The Market

According to the latest forecasts, cable's share of the domestic

TV industry will shrink from 91% at present to 77% by 2000. The number of DTH subscribers, after more than doubling to 5.5 million in 1996, could grow to almost 30 million globally, by the end of 2002.

The DTH satellite dish has emerged as the fastest selling product in the history of home entertainment. This has attracted the attention of telecommunications companies, who see the DTH satellite service as an important link in the Information Superhighway. AT&T has a 2.5% stake in DirecTV, with an option to raise its stake to 30% over 5 years.

Rival telecommunications carrier MCI, which has been taken over by WorldCom Inc., chose the direct route: an equal partnership with News Corp., in which MCI has already invested $2 billion. It is building a satellite business after paying $682 million for the license to the last remaining satellite broadcast spectrum that covers the entire United States. The partners expect to invest an additional $1 billion in the short term. The system will be called American Sky Broadcasting (ASkyB).

The MCI/News Corp. partnership has said its satellite capacity will be split more or less equally between home entertainment and business. The service will feature a wide range of products including some yet to be invented.

Apart from access to the Internet, other uses identified in the recent link between MCI and the Microsoft software group include the transmission of technological data. The national Mayo Clinic healthcare chain already operates satellite links that allow the exchange of video, audio, and other data for "live" surgery and diagnostics.

The potential U.S. satellite TV market, at present limited to 60 million households because of the need for a clear line of sight between dish and satellite and other technical constraints, may see some consolidation over the next few years. The argument for consolidation is reinforced by improvements in digital compression technology and optical fiber feeds. These may be good enough within 3 years to enable cable operators to match what are currently satellite's main advantages for households: the ability to reproduce high-quality images and sound and provide more than 200 channels.

Launch Facilities

The growth in global satellite systems has created a shortage of launch facilities. There are more than 1,700 satellite launches planned or proposed in the next 10 years. Unfortunately, rocket launch reliability is almost as bad as it was at the dawn of the space age. On August 12, 1998, a 20-story Titan 4 rocket carrying a spy satellite blew up 40 seconds after liftoff from Cape Canaveral, Florida. Less than a month later, a Ukranian rocket crashed shortly after launch, and destroyed 12 satellites, valued at $160 million. These satellites were part of Global Star's new 48-satellite PCS network.

With roughly 1 in 10 launches failing, insurance premiums are skyrocketing. The problem is more acute for the large GEO satellites. There are only six cities in the world that can launch these satellites. Roughly 60% of that business is handled by Europe's Ariane space consortium. About 30% of the launches fall to McDonnell Douglas Corp. and Lockheed Martin and Harris Corp., which use Delta and Atlas rockets from Cape Canaveral and Vandenburg Air Force Base. The remainder goes to Russian, Ukranian, and Chinese programs.

The smaller LEO satellites do not need the powerful Ariane-style rockets because their orbit is closer to Earth. Instead, these satellite operators are looking for unconventional launch schemes. Orbital Sciences Corp. of Dulles, Viginia, avoids the wait for ground launch openings by firing its Pegasus rockets from under the wings of a jumbo jet at 40,000 feet, for example. Sea Launch Group, a joint venture by Boeing Co. and Norwegian, Russian, and Ukranian partners, launches rockets from a floating oil rig in the Pacific Ocean. The converted platform can move from its home-port in Long Beach, California, to the equator for the most efficient geosynchronous launches, or steam a few hundred miles off the California coast for launches of MEO and LEO satellites. Sea Launch is sold out for its first 3 years of operation.

Conclusion

Since their introduction in the 1960s, satellites have been used in the commercial area primarily to broadcast simple one-way television and radio signals and to transmit two-way voice telecommunications traffic over long distances.

However, regulatory changes and continuing technology advances have highlighted the advantages of satellites in delivering global telecommunications services to all areas of the world.

The satellite business could more than triple to $29 billion by 2000 as big aerospace players and dozens of new startup companies race to offer new consumer services.

Satellites will provide a key link in the global Information Superhighway. They will extend the reach of terrestrial networks and complement wire line technologies by filling in the gaps.

Most geostationary orbiting satellite footprints, however, traverse national boundaries. This creates financial opportunities to increase traffic, but it also creates difficult political and cultural challenges. Despite the potential for satellites to operate without borders, national governments still attempt to impose border limitations by denying "landing rights" for foreign satellites and restricting the amount of foreign programming available to national cable television, broadcast, or DBS operators.

As a result, the FCC does not permit DBS operators in United States to use Canadian satellites, and Canada does not allow DBS operators in United States to beam their direct-to-home service into Canada. This has created a huge gray market in Canada for the 18-inch dishes. Canadians buy the dishes in United States, bring them across the border, and use an American post office box number to pay for the services.

However, a recent Canada/U.S. agreement will end Telesat Canada's monopoly on domestic satellite service by March 1, 2000. In addition, Canada will allow phone companies to use foreign satellites to offer Canadian telephone services (but not DBS services) by the same date.

The transborder nature of a satellite footprint all but eliminates mutually exclusive domestic and international markets. National governments will find it impossible to prohibit or regulate the use of satellite terminals and the extent to which citizens have access to the rest of the world.

Bibliography

Nellist, John G., *Understanding Telecommunications and Lightwave Systems,* New York, NY: IEEE Press, 1996.

"The Satellite Biz Blasts Off," *Business Week.* January 27, 1997 p. 63.

7 The On-Ramps to the Information Superhighway

Airplanes are interesting toys, but of no military value.

— Ferdinand Foch,
French military commander, 1911.

A Brief History

For most of the 21st century, access to both local and long distance telephone service in the United States has been provided by a Bell operating company or an independent telephone company. Each was a regulated monopoly in its service area. In Canada, access has been provided by the provincial telephone companies. In Europe and many countries around the world, telecommunications access has been provided by the government operated Posts, Telephone and Telegraph (PTT) organizations.

With the exception of telecommunications connections made by entities such as railroads and power utilities through their privately owned networks, every premise in the world equipped for telephony was connected for local service through wire pairs to a local switching center, commonly called a central office. This connection is known as the local loop. Private networks also used the public switched telephone network (PSTN) to back up and to extend their systems. Long distance connection to the worldwide network was made by switched connections to long distance carriers called interexchange carriers. The bandwidth of these connections was 300 to 3,000 Hz with occasionally noisy lines and with other line impairments that made data transmission difficult. Specially conditioned analog lines for data transmission could be leased for an extra charge. Modems that transmitted data at rates of 1,200, 2,400, 4,800, or 9,600 bps could be leased from the telephone company.

Distribution Services

Broadcast television and radio, cable television, and satellite television are distribution services that do not allow the user to select the type of information received nor to control the start and order of the presentation. For over half of this century, these services have provided users with access to informational services such as the latest news and weather and traffic reports, as well as entertainment. Because they are wireless, they can be accessed by users who are in motion as well as by those who are fixed.

Local Telephone Companies

Today's access network in North America is still mainly dependent on connections between the customer's premises and the central office of a local telephone company, called local exchange carriers (LECs) in the United States. At the central office, as shown in Figure 7-1, the local loop line from one premise may be switched to other premises in the same serving area, to other central offices for connection through switches and local loops to premises in other serving areas, or to a toll center for connection to a long distance provider. In countries around the world, local access is provided by lines between the

Figure 7-1   Traditional Access Through Local Telephone Company.

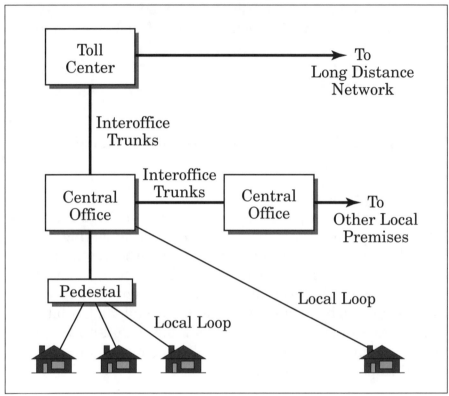

user premises and a local office of that country's PTT organization.

By 1995, 94% of all households in the United States and 99% of all Canadian households had wireline telephone service with a total of 150 million lines installed. Worldwide, there are more than 700 million lines that serve over a billion users. These lines make up the on-ramps by which users reach the Information Superhighway. Many homes have multiple lines for facsimile transmission and reception and for connection of computers with modems. Approximately 15% of the telephone lines in the United States utilize PCM digital loop carriers (discussed in Chapter 1) between the central office and a remote terminal. The analog signals on the telephone lines are converted into digital pulse trains, which are multiplexed together at the central office and are demultiplexed at the remote terminal, converted back to analog signals, and distributed over copper pairs to the premises.

The Public Switched Telephone Network (PSTN)

PSTN still provides much of the interactive services access to the Information Superhighway for users at fixed locations. Internet users dial numbers within their local calling area to reach Internet service providers (ISPs) without toll charges. New modulation techniques and encoding schemes have been devised to increase the number of bits represented by each symbol transmitted by an analog modem, allowing higher bit rates to be sent within the bandwidth available on dial-up telephone lines. Low-priced modems that will transmit at data rates up to 33.6 Kbps are common today.

Dedicated Analog Data Circuits

Analog channels with a bandwidth from 300 to 3,000 Hz are still being offered by telephone companies. Modems are required to transmit data over these lines at data rates up to 33.6 Kbps. Voice can also be transmitted over these channels. Because the channels are not allowed to use the exchange and toll network, they are of limited value in accessing the Information Superhighway. Their main value is the ability to control the security of transmissions over them.

Switched 56-Kbps Service

The telephone companies began offering switched 56-Kbps digital service in the mid 1980s. It is still offered today, but it is more expensive than ISDN. Data can be transmitted nearly six times as fast using this service, than it can be through a 9,600-baud modem and an analog telephone line. This service is useful for transmission of compressed signals of limited-motion videoconferencing telephones.

A standard 10-digit telephone number is provided for switched 56 digital service (SDS56) terminations, and local, long distance, and international numbers may be dialed. SDS56 is offered with a nominal monthly charge for the termination and the same usage charges as a voice channel.

Integrated Services Digital Network (ISDN)

ISDN was conceived in the late 1970s for digital service direct to the subscriber premises. It was not implemented on a large scale until the early 1990s. Basic rate ISDN service, with a total transmission rate of 144 Kbps, is offered with two bearer channels (called B channels), each of which can carry digitized voice, video, or data information at a rate of 64 Kbps, and one D channel, which carries control information and optional packet data transmission, as shown in Figure 7-2. Primary rate ISDN is a 1.536-Mbps service that provides 23 B channels at data rates of 64 Kbps each, plus an additional 64-Kbps D channel for control information, and one directory number is assigned to each ISDN B channel. A typical user access model with definitions of terminations and reference points is shown in Figure 7-3.

Figure 7-2   ISDN Basic and Primary Interfaces.

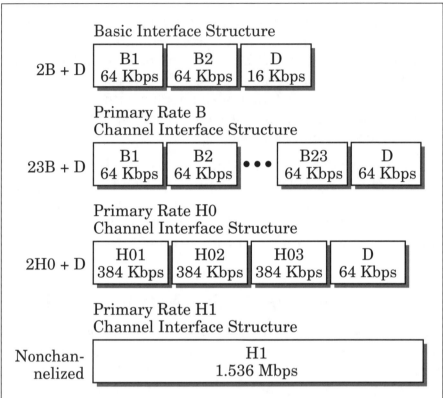

Leased High-Capacity Digital Lines

The telephone companies also began offering leased DS1, DS3, and fractional DS1 (256, 384, 521, or 768 Kbps) channels at digital transmission rates of 384 Kbps and above. Using the new compression algorithms, full-motion video picture quality is very good and throughput for data circuits is high. DS1 and fractional DS1 services are cost effective when many lines are required at one location. In this situation, a private branch exchange (PBX) switch equipped with a DS1 interface is usu-

Figure 7-3   Typical ISDN User Access Model.

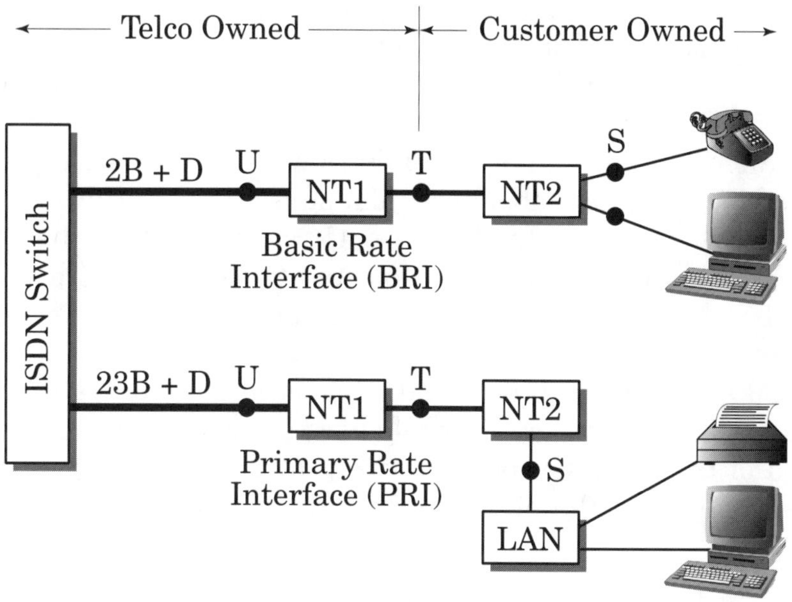

LAN – Local Area Network
NT1 – Network Terminal 1 (terminates transmission line)
NT2 – Network Terminal 2 (connection for multiple terminals)
S   – Switched Reference Point
T   – Terminal Reference Point (customer interface)
U   – Reference Point (transmission line to local switch)

ally provided to switch the lines internally. For locations such as business, manufacturing, and government facilities, where upwards of 100 lines are required, DS3 service with a DS1 to DS3 multiplexer will probably be cost effective. One DS1 circuit, which has the capacity of 24 telephone lines, can be leased for the cost of eight private telephone lines. Twenty-eight DS1 lines can be multiplexed into one DS3 line, which can be leased at the price of eight DS1 lines.

Cellular Radio Providers

Persons in vehicles and on foot have access to the global telephone network through cellular radios. In 1991, there were only 7.6 million cellular telephone subscribers in the United States. This number increased to 55 million by the end of 1997. In Canada by the end of 1997, there were 4.2 million cellular subscribers. The number of cellular subscribers worldwide has been growing at a rate of 35% to 45% per year over the past few years, and this trend is expected to continue. The total number of cellular subscribers in the entire world passed the 100 million mark in 1996 and is projected to reach 200 million in 1999 and 284 million in 2001. In the United States and Europe cellular systems initially used analog technology. One third of the 5 million persons living in Finland carry mobile telephones, and the number is increasing at the rate of 27% per year.

Three cellular radio standards prevail throughout the world. The details of these standards are discussed in Chapter 9. In the United States, the Advanced Mobile Phone System (AMPS) technology was initially used. In Europe, problems developed with incompatibility between systems in the various countries and a new standard called Global System for Mobile Communications (GSM) was developed. The GSM systems facilitated international roaming of cellular stations so that communications could be maintained even when crossing international boundaries. GSM was first installed in 1991, and today GSM is used throughout much of the world. In North America, many of the cellular operators are upgrading their facilities to digital transmission with systems that are derivatives of GSM.

Specialized Common Carriers

Until 1972, when the FCC began licensing specialized common carriers (SCCs) such as Sprint and MCI, almost all (97%) of the long distance communications services in the United States were provided by the long lines of AT&T. Today, there are over 50 SCCs, including Sprint, MCI, and WorldCom, sharing long distance revenue with AT&T. Connection from subscriber premises to an SCC in many instances is currently provided by the local service provider. The SCCs have established points-of-presence (POPs) in most cities, in some cases in the offices of local service providers.

Private Networks

Many large companies and government organizations have built their own telecommunications networks. These systems consist of a mix of privately owned microwave radio systems, private lines leased from local and long distance common carriers, leased satellite channels, and connections to the PSTN.

Railroads, power utilities, oil and gas pipelines, and state and municipal government agencies that own right-of-way, have installed fiber optic networks to supplement their microwave radio systems. Many of the railroads have leased transmission channels on their fiber optic networks to common carriers to use as part of their networks. In other arrangements, they have leased so-called dark fibers to which the carriers attach their own lightwave transmission equipment.

Bypass Carriers

An FCC ruling in the mid-1980s gave the power utilities permission to lease excess capacity of their systems to other telecommunications users. The rates offered by these companies for leasing DS1 and DS3 channels were usually much less than those charged by the local telephone companies. The high-capacity channels are usually used for transport of voice, video, and data between facilities within a local calling area or to connect the users with an SCC POP, bypassing the networks and charges of the RBOCs and other local service providers. For this reason, the entities providing the service became known as bypass carriers.

Distribution Services

Access to distribution services, such as dial-in information services and Teletext, which allow users individual presentation control, is typically provided through the PSTN.

Broadcast distribution services provide entertainment, education, and information. Ninety-nine percent of the homes in the United States have radio receivers, with an average of 5.6 radios per home. More than 99% of automobiles in the United States are equipped with radio receivers, which provide mobile listeners with continually updated news, information on traffic and weather (including warnings of hazardous road conditions, blocked roads, and impending storms), in addition to educational programs and entertainment.

Broadcast, satellite, and cable television provide residential and commercial premises access to news, entertainment, educational, and informational television services. Ninety-eight point three percent of the homes in the United States receive broadcast television, with an average of 2.2 sets per home. Sixty-two point five percent of U.S. households are equipped with cable television. Small satellite dish receiving systems are being sold in consumer electronic stores, large retailers, and mass marketing warehouses at prices around $200 after rebates. Programming charges are about the same as those for cable TV. At the end of 1995 there were 2.23 million U.S. homes with satellite TV receiving systems. By the end of 1996, that number had more than doubled to over 5.5 million, which is more than 5% of U.S. households with television. While cable TV systems are being upgraded to provide optional interactive services, it is not feasible to provide two-way transmission for residential satellite TV.

On-Ramps of the Future

In the visible future, the more than 700 million lines that serve over a billion users worldwide will continue to be the on-ramp by which many of these users reach the Information Superhighway. The local telephone companies that will use these lines to compete to be access providers for much of the POTS and data transport are finding new ways of exploiting the existing infrastructure of copper pairs and digital loop carrier. They are also exploring new technology that will allow them to provide new and enhanced services to their subscribers. Most of the new technology will provide broadband services, such as television programs and high data rate digital communication, to compete with cable and wireless TV companies, in addition to access to PSTN. Some of the technology utilized by the LECs to provide local access is described in the following paragraphs.

High Data Rate Subscriber Line (HDSL)

HDSL uses existing copper in the local loop to provide DS1 service. The technology that is used to transmit at high data rates over unconditioned copper pairs is an outgrowth of ISDN. HDSL uses the unique ISDN line code, which has a symbol rate that is half the bit rate. The symbol (baud) rate is 392 kilo baud which, because of the coding, provides a bit rate of 784 Kbps. As shown in Figure 7-4, HDSL requires two loop pairs, using echo suppressors, with each pair operating at 784 Kbps in each direction. The signals from the pairs are multiplexed at each end to provide a DS1-compatible signal with a net data rate of 1.544 Mbps, making HDSL a repeaterless replacement for a T1 line.

Asymmetric Digital Subscriber Line (ADSL)

One of the technologies being used by telephone companies to exploit their embedded plant of plain old copper pairs is ADSL, a loop carrier system that transmits with a data rate up to 9 Mbps, depending on the distance, in one direction (to the premise). Maximum distance between the central office and the premise is 18,000 feet, which should reach most subscrib-

Figure 7-4   Digital Loop Carrier: HDSL.

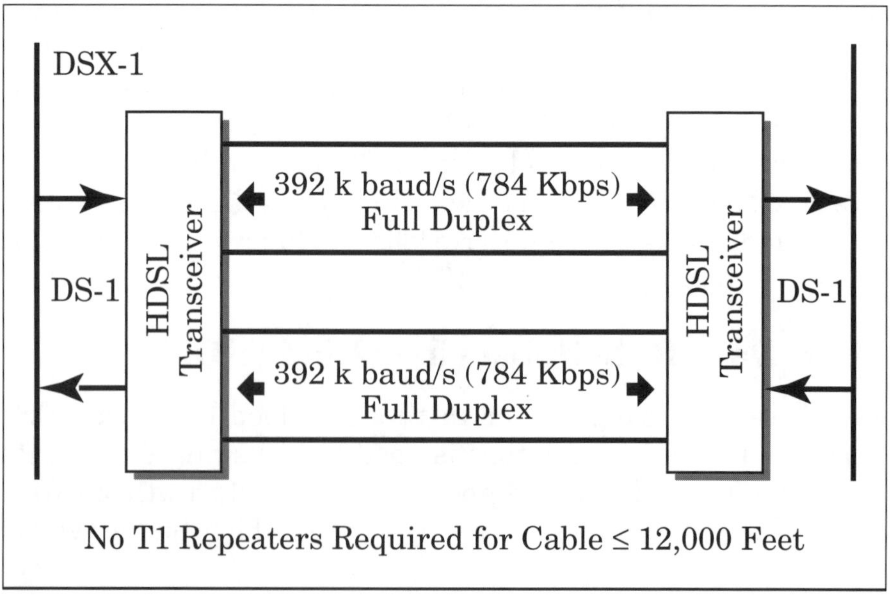

DSX-1

392 k baud/s (784 Kbps) Full Duplex

392 k baud/s (784 Kbps) Full Duplex

DS-1

HDSL Transceiver

HDSL Transceiver

DS-1

No T1 Repeaters Required for Cable ≤ 12,000 Feet

Figure 7-5   Asymmetric Digital Subscriber Line.

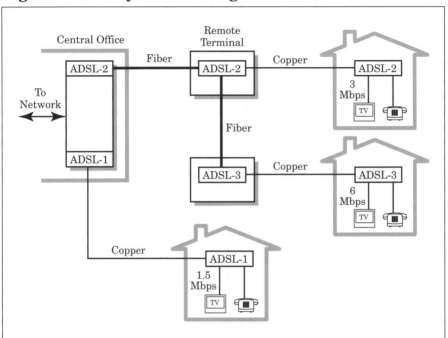

ers. ADSL, as illustrated in Figure 7-5, can deliver a 1.5-Mbps channel over a distance of 18,000 feet, a 3-Mbps channel over a span of 12,000 feet, or a 6-Mbps channel over loops up to 8,000 feet long. It also provides a POTS voice circuit in both directions, plus a low-speed (16 Kbps) digital maintenance and control channel. Maximum data rate versus maximum distance for various digital loop systems, including ADSL, are compared in Table 7-1. An advantage of ADSL over HDSL is that if a data rate between 1.544 Mbps and 9 Mbps is required in one direction only, it can be implemented with only one pair.

Applications of ADSL include transport of Motion Picture Experts Group 2 (MPEG-2) compressed video at the 1.544-Mbps rate. Another application is the transfer of high-speed digital data from a computer or server to the user, with much slower

Table 7-1   Digital Loop Systems.

Name	User Payload	Max. Distance from CO	Pairs Required
ADSL-1	1.5 Mbps	18,000 feet	1
ADSL-2	3 Mbps	12,000 feet	1
ADSL-3	6 Mbps	8,000 feet	1
Dedicated Digital Services (DDS)	2.4 to 64 Kbps	12,000 feet (more at the lower speeds)	2
Switched 56	56 Kbps	18,000 feet	1 or 2
ISDN DSL	144 Kbps	18,000 feet	1
Repeatered T1	1.536 Mbps	up to approximately 200 miles	2
HDSL	1.536 Mbps	12,000 feet	2

data transmission from the subscriber in the reverse direction, which would be useful for access to the Internet. Large files from the Internet could be downloaded at high speed, and the slower responses from the man-machine interface of the user could be transported at lower data rates to the Internet. One of the RBOCs has announced that it will provide Internet access over ADSL circuits.

An enhancement of ADSL, called rate adaptive digital subscriber line (RADSL), allows the downstream data rate to vary from 600 Kbps to 7 Mbps according to the signal carrying quality of the telephone line. The upstream rate can vary from 128 Kbps to 1 Mbps. In addition, RADSL can operate beyond the 18,000 feet of ADSL by lowering the transmitted data rate.

Fiber in the Loop (FITL)

Telephone companies have begun deploying FITL systems as an alternative to adding more copper in their distribution plant when additional transmission capacity is required. Figure 7-6 is an example of a typical copper distribution system. The copper subscriber carrier consists of the feeder to a remote terminal and the subfeeder to a distribution cabinet. The copper cables fan out from the cabinet and service access points are splices between the cable and drops to individual premises. In the evolution to a fiber distribution system illustrated in Figure 7-7, the remote terminal of the subscriber carrier is replaced by a host digital terminal (HDT). In Figure 7-8, fiber cables fan out from the HDT through a passive optical network (PON) to optical network units (ONUs), which can be considered the service access points.

The optical signal on each fiber of a sub-feeder cable is split and transported over multiple fibers to individual ONUs that

Figure 7-6   Typical Copper Distribution System.

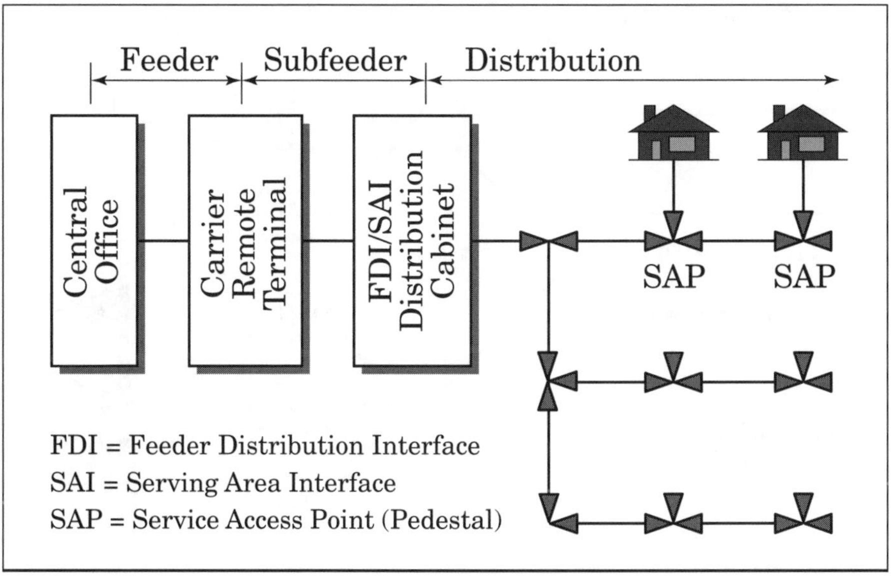

Figure 7-7   Typical Fiber Distribution System.

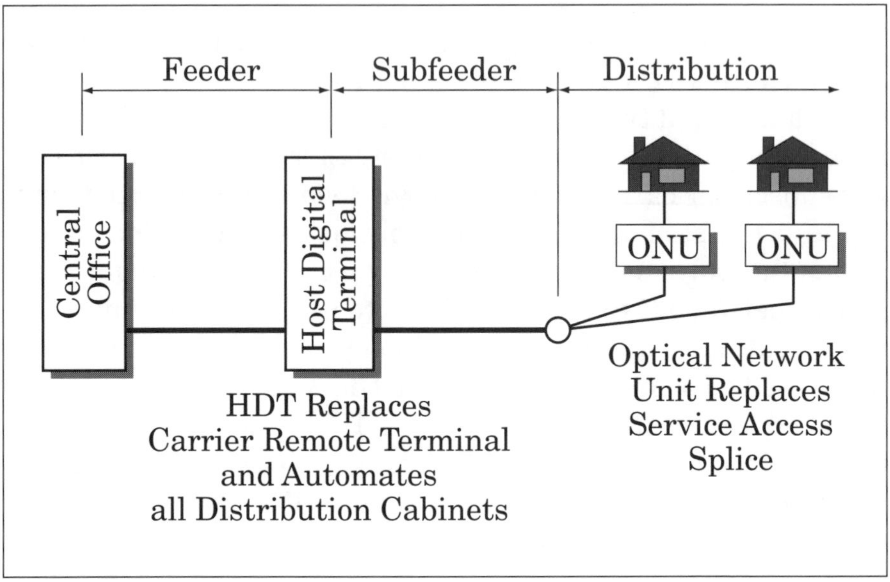

Feeder   Subfeeder   Distribution

Central Office

Host Digital Terminal

ONU   ONU

HDT Replaces
Carrier Remote Terminal
and Automates
all Distribution Cabinets

Optical Network
Unit Replaces
Service Access
Splice

Figure 7-8   FITL System with PON.

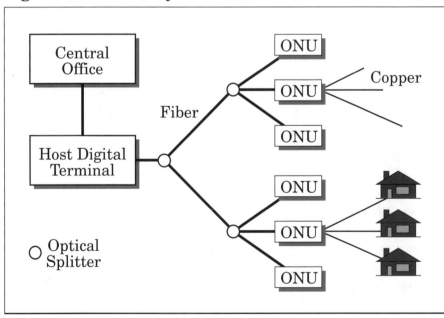

Central Office

Host Digital Terminal

Fiber

ONU

ONU   Copper

ONU

ONU

ONU

ONU

○ Optical Splitter

perform optical to electrical conversion. Each ONU is capable of providing up to 96 access lines. Transmission over FITL systems consists of a digital PCM signal from the central office to the ONU. Service from the ONU to the premise may be analog telephony, ISDN, or T1 digital lines. Copper pairs connect the optical network units to the premises. In the reverse direction, the ONU provides electrical-to-optical signal conversion.

Fiber-to-the-curb and fiber-to-the-premise video systems, which can be co-deployed with the POTS FITL systems, can also provide television access. A number of systems are on the market with different system architectures. Figure 7-9 is a diagram of a representative fiber-to-the-premise system that transports digitized voice channels and 24 channels of digitized compressed video from the HDT to an ONU. Two drops are installed from the ONU to network interface devices at each premise: copper twisted pair for POTS, and coaxial cable for video. The coaxial cable delivers three simultaneous digital video feeds to the premise where they may be terminated in up to three digital video terminals (DVTs). Each set-top DVT converts one of the feeds to a standard TV signal for use by a TV receiver. Each DVT may access any of the 64 channels transported from the HDT. The twisted copper pairs deliver POTS service.

A reverse channel from the DVT to the HDT allows interactive channel selection and pay-per-view services to be selected by a subscriber by means of a hand-held remote unit. The reverse channel also provides access to the Information Superhighway.

An FITL network in which a common fiber is shared by a video ONU and a telephony ONU to provide video-to-the-premise and POTS service is pictured in Figure 7-10. This system distributes standard, amplitude-modulated TV signals, so a set-top box is not required.

Figure 7-9   Digital FITL System.

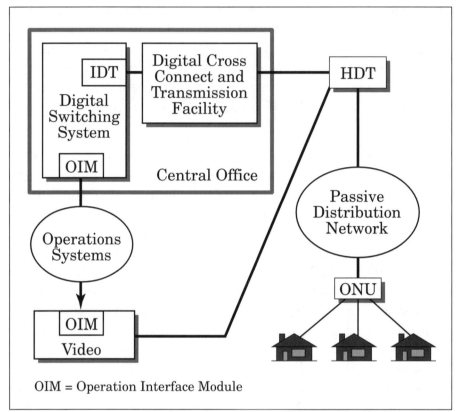

OIM = Operation Interface Module

Cable TV Companies

Cable TV companies have been providing broadcast distribution services for nearly half of this century, and today they reach 63% of the households with television. The original infrastructure, much of which is still in place, consisted of a receiving site, called the headend, where signals from television stations and satellites were connected to a trunk cable, and a branching network, which distributed the signals throughout the company's franchise area. The network amplified and distributed

Figure 7-10   Typical Co-Deployed POTS and Video System.

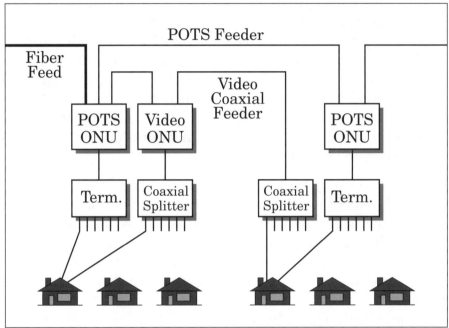

the analog TV signals that were transmitted by the TV broadcast stations, and these signals could be accepted directly by the subscribers' television sets. These coaxial cable networks typically have a total available bandwidth between 335 and 375 MHz and are capable of transporting up to 50 television channels. Frequencies between 5 and 40 MHz are allocated to transmission in the return direction (from the user to the headend), but most of the existing infrastructure is presently not equipped to transport signals in the return direction. Additionally, the return paths of cable TV systems that have been activated may not be robust enough without extensive upgrading to provide the reliability and availability that users have become accustomed to experiencing with PSTN.

Hybrid Fiber/Coax Systems (H-F/C)

Though much of the existing coaxial cable TV plant cannot support high speed data transmission over the return path, CATV system operators are busy rebuilding 12% to 15% of their installed plant annually with H-F/C architecture and activating and upgrading the return channels at the same time. By 1999 cable networks will have been rebuilt so that 80% of their subscribers will be served by upgraded systems. Some telephone companies are also installing H-F/C networks and offering analog video service to their customers.

A diagram of a typical H-F/C system is presented in Figure 7-11. Signals of the services to be provided are converted by radio frequency modems to frequencies between 50 to 750 MHz and combined into a composite radio frequency signal. The light output of an optical transmitter is modulated with the resulting signal and transmitted downstream over optical fibers to optical/electrical nodes in the vicinity of the subscribers. An upstream path with a bandwidth typically from 5 to 40 or 50 MHz provides transmission of data and voice signals in the reverse direction.

Downstream optical signals are converted to radio frequency electrical signals at the optical/electrical nodes, and coaxial cables transport the downstream signals from the nodes to network interface units (NIUs) on the outside of each premise. The NIU separates the signals, and additional cables and copper pairs connect the NIU to the TV, the telephone, and the computer. A set-top converter is required at the TV set to translate the signals to standard HF and UHF television channels. Proposed allocation of spectrum within the 5 to 750 MHz bandwidth of the H-F/C systems is shown in Figure 7-12.

Figure 7-11   Typical H-F/C System.

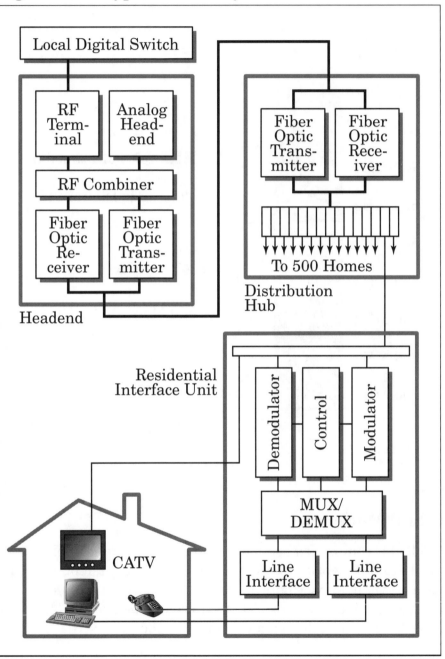

145

Figure 7-12 Proposed HF/C Spectrum Allocation.

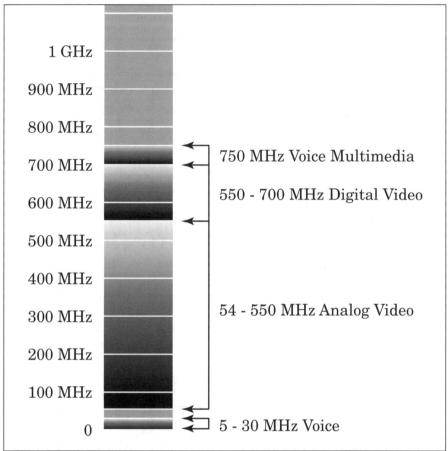

CATV Modems

Cable TV companies are using cable modems and their existing H-F/C networks to offer broadband video, telephony, and
data to their subscribers in competition with other local exchange providers (Figure 7-13). These modems and H-F/C provide 80 or more television channels in the downstream direc-

Figure 7-13   Internet Access with CATV Modem.

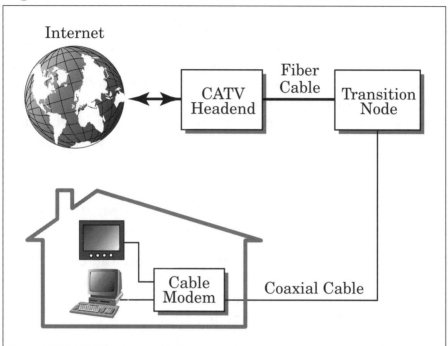

tion plus telephony, and data transmission at rates from 4 to 10 Mbps in the upstream and downstream directions. Using the modems, cable companies can provide Internet access and other data transport at rates 1,000 times those of PTSN.

Cable TV companies and vendors have agreed to work together to generate standards for interoperable first-generation modems. The cost of each modem is $200 to $450 in large quantities, depending on the manufacturer. Over 10 million modems have been ordered by cable TV companies in 1996. And high-speed Internet access is being offered in numerous cities at prices from $30 to $45 per month.

Competitive Access Providers (CAPs)

In the United States the Telecommunications Act of 1996 has opened the door for LECs (local telephone companies) to provide long distance, cable TV, and other telecommunications services. New companies, as well as existing CATV, cellular, and long distance providers, are now allowed to compete with established local telephone companies—now called incumbent local exchange carriers (I-LECs)—for local telephone service revenue.

The Telecommunications Act also allows RBOCs to offer local access service to customers of adjacent independent telephone companies as alternative local exchange carriers.

One of the issues that must be resolved before there will be an incentive for users to change carriers is how to provide local number portability, so that users will not be required to change telephone numbers if they change carriers.

In Europe and the United Kingdom, licenses have been issued to CAPs, which are competing with the public networks. In many instances, the CAPs are seeking to woo the users by offering service at lower rates. In the United Kingdom, for instance, some CAPs offer telephone services at prices 10% less than British Telecom. In Germany, rates 15% or more below the Deutsche Telekom prices are offered.

MCI, WorldCom, British Telecom, AT&T, Sprint, and other long distance carriers are free to compete in local access markets. MCI agreed in 1997 to be acquired by WorldCom in an exchange of shares, to form the second largest global telecommunications company. The new company, in addition to being an international long distance mega-carrier, is expected to become a major player as a competitive local exchange carrier (CLEC) in local access markets in the United States and Europe. It has contracted to build a fiber optic network that will link London, Amsterdam, Brussels, and Paris, and connect with the Gemini network, which connects London and New York.

Multichannel Multipoint Distribution System (MMDS)

MMDS was first used in North America in Manitoba, Canada, in 1996. Another North American demonstration of MMDS took place in mid-1996 when Pacific Bell showed off its digital system that broadcasts MPEG-2 compressed video from the top of Mount Wilson in the Los Angeles area. Another antenna is located on Mount Modjeska to serve the Orange County area. Today, the system is capable of offering 136 channels of MPEG-2 video (including 40 channels of near video-on-demand) and two-way high speed Internet access. Both the downstream and upstream transmission paths will use wireless cable frequencies in the 2.1- and 2.7-GHz bands. By using sectorized receiving antenna patterns, which divide the circular reception pattern into 7.5-degree sectors and use two frequencies in the upstream direction and alternating frequencies between adjacent sectors, Internet service can be offered to 100,000 subscribers from each antenna location. Pacific Bell expects to have access to 2.5 million subscribers in the Los Angeles, Victorville, and San Francisco areas. Additionally, Pacific Bell purchased MMDS licenses in 11 other metropolitan areas during the FCC auction of channels. Several of the RBOCs are considering the use of MMDS as an interim solution to entering some of their available markets quickly. The RBOCs, Bell Atlantic, and Nynex have reconsidered their original plans to install MMDS because of the wide availability of fiber-based networks.

Local Multipoint Distribution System (LMDS)

LMDS is a wireless interactive communication system that can deliver 50 analog or 200 digital TV channels plus high-speed digital access to the Information Superhighway in the return direction. Because it operates at 28 GHz in the millimeter wave region, LMDS can take advantage of reflections, or bounces, and is not limited to line-of-sight operation as is MMDS which operates at lower frequencies. LMDS is being tested at numerous sites around the world. In the United States, the FCC has allocated 1.3 GHz of spectrum in the 27.5- to 29.25-GHz frequency band for LMDS. The FCC auctioned entire 1.3-GHz slices of spectrum to single licensees in each of 493 basic trading areas in the United States and its territories. In Canada, the government is licensing 1 GHz of spectrum in the 28-GHz region for what it calls local multipoint communications systems (LMCS), and has plans to add another 2 GHz within the next 2 years. Equipment has been developed that will support two-way T1 (1.544 Mbps) data as well as 200 digital TV channels.

To connect with an LMDS network, a user needs a rooftop or windowsill antenna and an LMDS digital converter, which connects directly to a television set and to the telecommunications terminal equipment. The present cost of the converter is $450. The price is expected to drop quickly to about $300.

In the United States, the objective of the FCC in licensing the LMDS service is to stimulate competition to the CATV operators, believing that cable and telephone companies dominate their markets without yet offering much competition to each other. It is expected that LMDS will be a third competetive force in broadband communications. It may also be an opportunity for newcomers, smaller entreprenurial firms that do not have infrastructure in place as do telcos and CATV operators, to get into the market.

PCS and Cellular Radio

Wireless communication subscribers worldwide now make up 10% of the installed fixed lines, and, currently more mobile telephone numbers are being issued than fixed.

PCS will offer digital wireless access to the Information Superhighway in competition with the incumbent cellular radio operators. Although the primary use will be wireless telephony, PCS will offer a wide variety of new applications such as two-way paging, voice messaging, and e-mail.

PCS will utilize smaller sized cells than the existing cellular system and it is expected to provide enhanced indoor coverage. One application of PCS may be the replacement of fixed line telephone service within the premise. PCS cellular can replace fixed telephones and cordless portable telephones within premises and become a mobile cellular telephone away from the premise. In a residence, instead of buying a telephone for every room, a PCS unit can be purchased for each member of the household. A lifetime telephone number assigned to each person will make a person reachable anywhere.

The Wireless Local Loop

Cellular radio and PCS can also be used as alternatives to the installation of new cables by providing a wireless local loop for access to the Superhighway in countries with inadequate telecommunications networks. Developing countries need to improve telecommunications to their rural populations. In most of the developed countries, there is an average number of 50 telephones per 100 persons. In some countries in Africa, the average is less than 0.5 per 100. In Brazil, 91% of the rural population, 81% of the residences, and 47% of the businesses do not have any telephone service. Studies have shown that a 1% increase in the number of telephones available in a country can result in a 3% increase in per-capita income.

Several factors discourage the upgrading and extending of existing systems. The long distances and difficult terrain in many of the rural areas make the installation of wires and cables difficult and expensive. Funds for such projects are limited in many of the countries. These problems can be solved in many instances by providing radio communications, particularly cellular technology, which can be deployed less expensively and more rapidly, and is less sensitive to terrain difficulties than wire and cable systems.

Because it uses advanced cellular technology, the wireless local loop can provide data and facsimile communications in addition to voice. It can also provide communications in times of emergency and can be used to back up existing land line networks.

In the United States, long distance carriers want to enter local markets as competitive local exchange carriers but do not have the infrastructure in place to accomplish this. One option is to lease local telephone lines from the incumbent local exchange carriers at a discount and resell access service to users. Another approach is to use wireless technology to bypass the I-LECs' network, and to avoid the access fees. AT&T is considering a system that would let subscribers use the same handset to call fom home, the office, or the car for a flat monthly fee only slightly higher than existing phone rates. MCI is planning to provide wireless access through its venture with NextWave, which owns wireless licenses in 63 markets. Sprint

PCS, a joint venture of Sprint, TCI, Comcast, and Cox Communications, is also considering wireless as a way to provide local access. Winstar Communications in New York uses radio in the 38-GHz band to offer local access, including higher bandwidth channels, at lower prices than the I-LECs. Winstar holds 38-GHz licenses in 46 of the top U.S. markets.

Paging Systems

Initially, pagers beeped or vibrated to let wearers know that they were being paged so that they could call the office or home to receive their message. This was a kind of one-way ramp to the Information Superhighway. Recently, two-way, or acknowledgment paging, was introduced. This technology allows a subscriber to receive a few lines of text on a pager and to reply to the sender of the message with a selection from a customized list of responses. Software is available that will allow a message typed into a modem-equipped personal computer to be sent over a paging system and received on a pager. Some pagers can store up to 20 messages.

Today the subscriber is able to receive a letter on a wireless messaging device and respond with either a written or voice message. Pagers are approaching digital cellular radio and PCS functionally, but the amount of air time used is much shorter than a cellular conversation.

GEO/LEO/MEO Satellites

Only 15% of the world's land mass receives coverage by cellular telephone networks. Global access to the Information Superhighway outside the range of cellular radio will be provided to roaming users by universal PCS companies through geostationary Earth orbit (GEO), low Earth orbit (LEO), and medium Earth orbit (MEO) satellite systems. Sophisticated U.S. technology companies are pushing forward in the development and implementation of satellite-based phone systems that will enable telephone calls to be originated and received any place on Earth. The technical requirements and the economic considerations for these satellite systems are discussed in Chapters 6 and 9.

The GEO Inmarsat-3 satellites, launched for Comsat Corporation in late 1996 and early 1997, provided the first service for personal satellite terminals. Four satellites are required to cover the entire Earth. The personal satellite terminals are smaller than a notebook and weigh 5.5 pounds. For computer integration, e-mail, and Internet access, the unit plugs into a computer's modem port.

Each terminal user receives a subscriber identity module (SIM), a smart card that contains authorization, personal, and billing information. The system has global roaming ability that will allow the user to receive telecommunications services anywhere, through any terminal, using a unique universal personal telecommunications (UPT) "follow me" telephone number. Subscribers are charged $3 per minute for calls from anywhere to anywhere.

One of the first users of the system was a group that attempted to climb Mount Lhotse in Nepal. During their month-long expedition they were able to keep in touch with their families and friends, making over 25 calls.

Other companies are planning LEO and MEO networks to provide mobile telephone service to users with hand-held terminals. The LEO systems will use large numbers of satellites and earth stations to provide full global coverage. MEO systems can offer global coverage with fewer satellites and ground stations.

Motorola began launching the LEO Iridium satellites in early 1997 and began worldwide service late in 1998. Motorola has also filed plans for a second system with 72 LEO satellites. This system, to be called M-Star, will be aimed at business users and will be designed to transmit huge quantities of data at very high speeds. Service charges for calls for all of the systems range from less than $1 per minute to $3 per minute.

Global Positioning System

The Global Positioning System (GPS) consists of 24 satellites deployed by the U.S. Department of Defense in 1993. Four satellites are located in each of six planes, as shown in Figure 7-14. The satellites broadcast ranging codes that enable GPS receivers to determine the transit time of the transmissions and calculate the distance between a satellite and the receiver to an accuracy of less than 100m. By using a receiver in a known location and comparing the error between calculated distance and the known distance, a correction factor can be derived and the accuracy of the position calculated to within 1m or less. This type of system is called a Differential Global Positioning system (DGPS). By using information from several of the satellites simultaneously, three-dimension (latitude, longitude, and altitude) positional information can be available 24 hours a day, anywhere on Earth.

Many applications are being developed for the information that is available from the GPS system. Hand held receivers are available on commercial markets at prices around $100 for hikers and weekend sailors. Aircraft equipped with a Satellite Landing System (SLS), developed by Honeywell, have been used by the Federal Aviation Administration (FAA) to demonstrate DGPS-guided precision airport approaches and automatic landings. According to Honeywell, the DGPS systems can provide

Figure 7-14 Global Positioning System.

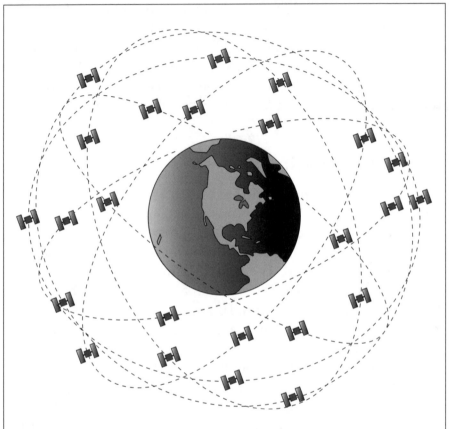

the highest level of en route and approach navigation accuracy available today, and, when coupled with an aircraft's flight management and flight guidance systems, can fly instrument approaches to virtually any airport runway in the world. GPS systems have been combined with cellular handsets so that police, fire, or other emergency services can be summoned to the phone's location at the touch of a button.

The Ford Motor Company and General Motors Corporation (GM) offer as optional accessories to their Lincoln Continental and Cadillac cars, GPS systems integrated with cellular tele-phones that provide wireless information and communications services.

The Cadillac accessory, named OnStar, can be activated by either of two buttons on the phone keypad, one for emergency service and the other for real-time services such as route infor-mation and direction to hotels and restaurants. The vehicle location, determined by the GPS system to an accuracy within 100m is transmitted over the telephone network to the OnStar center, which is manned by a staff of advisors 24 hours a day, 365 days a year.

The Lincoln Remote Emergency Satellite Rescue Unit (RESCU), also activated by a push button, dials an emergency or assis-tance number, sends the vehicle location over the cellular phone, and connects the driver to the assistance center.

Conclusion

Much of the interactive access to the Information Superhigh-way is still traditional POTS provided by local telephone companies over copper wire pairs, or by cellular telephones. The Telecommunications Act of 1996 has opened the way for new players to enter the market for provision of last mile serv-ices (i.e., access from the premises to the backbone lanes of the Superhighway). Local and long distance carriers, wireless, cellular, and cable TV operators are going after each other's markets. Narrow- band slow-speed access using the local access providers' embedded infrastructure has been provided in the past. The increasingly widespread use of the Internet and the desire of Internet users to download graphics, pictures, and text in a reasonable length of time is pushing the require-ment for higher speed access from the premise to the Superhighway. New competitors entering the market will offer higher data speeds and bandwidth using technology that is just emerging from the development stage. Some of the new wireless technology, such as MMDS and LMDS, is viewed by some of the telephone companies as an interim strategy for providing video. It will be limited in the number of users it can serve because the realizable bandwidth will be restricted by the scarcity of available radio spectrum. Cable TV companies, expecting to lose market share to direct broadcast residential satellites and to MMDS, are rebuilding their networks to compete for the local access market for POTS, data, and interactive video.

Optical fiber technology can transport terabits per second (millions of megabits per second) of information by a combination of higher data switching rates andWDM to provide the needed bandwidth. Local telephone companies and cable TV companies are installing fiber backbones within cities at a tremendous rate. Cable TV companies and telephone companies are beginning to extend these backbones and trunk systems to the premise: the cable TV companies with H-F/C, and the telephone companies with some digital and analog FITL, ADSL, and H-F/C. The Southern New England Telephone Company (SNET) is building a state-wide network to serve 1.2 million households, using nearly 20,000 miles of H-F/C plant at a cost of $4.5 billion.

The cable TV companies are executing massive programs to upgrade the upstream paths of their networks. They have ordered millions of modems, each of which will deliver over 100 video channels plus telephony downstream to the premise and multi-megabit data upstream from the premise to the Superhighway. Time Warner Cable (TWC), the second largest MSO in the United States with 11.7 million subscribers, deployed optical fiber backbones in its franchises 3 years ago and has begun a program to upgrade and extend its systems with H-F/C networks and cable modems. It plans to finish the upgrade by the year 2000 at a cost between $3.5 and $4 billion, equal to approximately $180 per home.

The result of the advances in telecommunications technology and the increased competition between access providers and between access technologies present consumers with a wide choice of on-ramps to the Information Superhighway. One of the benefits of competition for the consumer will be competitive pricing for the services. One telephone company reports that, when they began marketing cable TV and data services in an area, the incumbent cable TV operator offered discounted service and other incentives to its subscribers.

Improved technology will continue to be developed for both wired (copper and optical fiber) and wireless access to telecommunications, improving the availability of information and hopefully the quality and enjoyment of our lives.

It is expected that access networks will evolve in the very near future to a multi-tiered configuration of counter-rotating optical fiber rings that transmit data at multi-gigabit SONET/SDH digital rates. The lower rings in the tier will terminate in hubs from which connection to user premises will be made. An example of one step of this evolution is shown in Figure 7-15. Initially, a coaxial cable bus or a tree and branch architecture will be used for the final distribution to the premises. Eventually, the coaxial cable will be superseded by optical fiber to the premise.

This chapter describes the many pathways through which information can flow into and out of the Superhighway. With the many changes taking place (the deregulation of communications networks throughout the world, the agreements of the World Trade Organization (WTO) countries to allow foreign organizations to provide communications services within their

Figure 7-15   Network Evolution to OC-192.

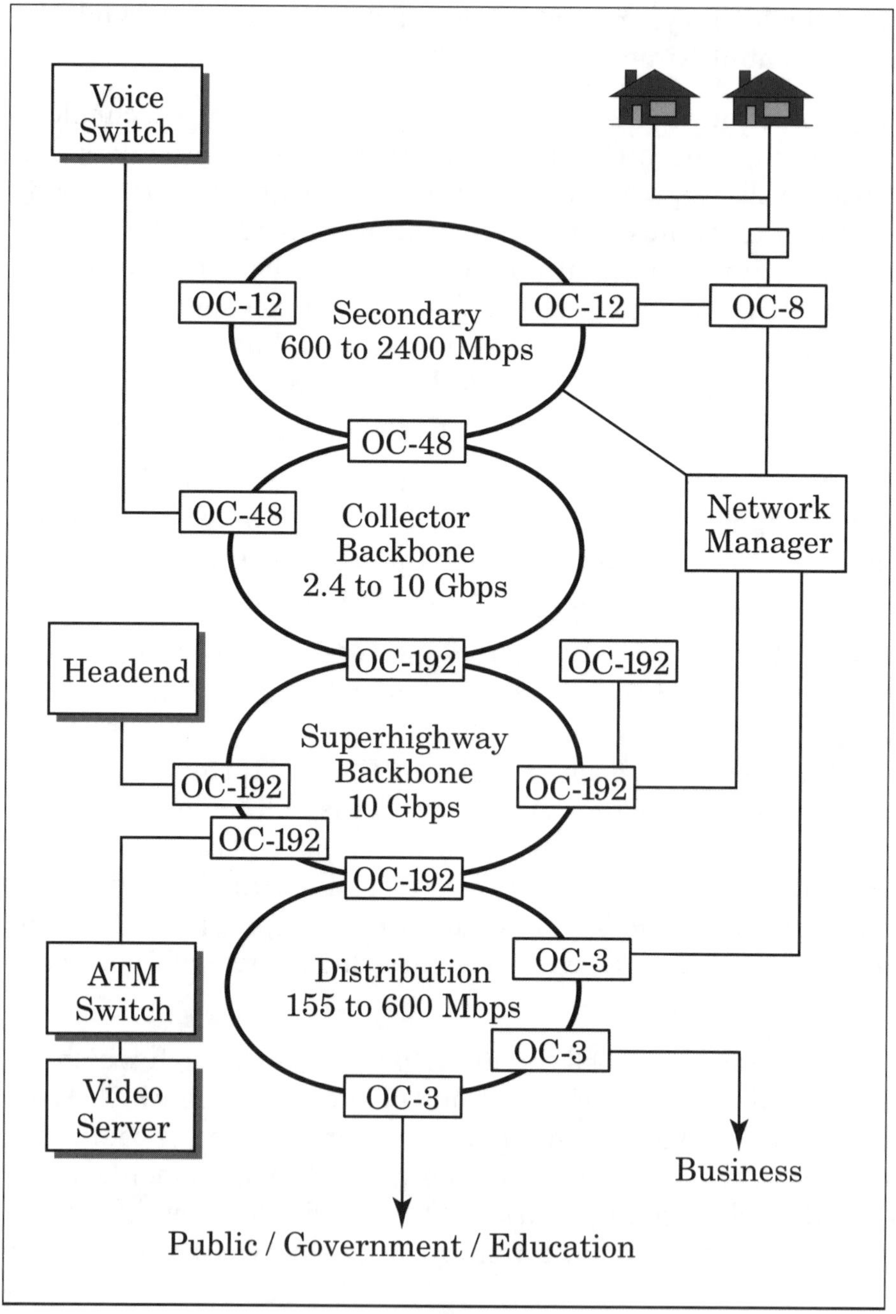

boundaries, the explosive development and application of new telecommunications technology, and the mergers and acquisitions of the communications providers), this chapter becomes a snapshot of the on-ramps as they exist at this time.

Bibliography

Brown, Roger, Cervenka, Dana and Lafferty, Michael, "Telco Video Plans Becoming Clearer All the Time," *Communications Engineering & Design,* November 1996.

Finneran, Michael, "Handicapping Competitors for the Local Loop," *Business Communications Review,* November 1996.

Gifford, Joe, "Wireless Local Loop Applications in the Global Environment," *Telecommunications,* July 1995.

International Telecommunication Union Press Service, Universal Personal Communications, *Microwave Journal*, July 1996.

Kennedy, Traver H., "Expanding the Communications Horizon," *Wireless Integration,* Vol. 1, No. 1.

Ramaswamy, Arun, "Delving Into Digital Compression," *Broadband Systems & Design,* December 1996.

"Satellite Phone Aids Mt. Lhotse Snowboarders," *Broadband Systems & Design,* December 1996.

Shankar, Bhawani, "GSM: The Upwardly Mobile Standard," *Microwave Journal,* July 1996.

U.S. Department of Commerce, *Statistical Abstract of the United States 1995,* U.S. Government Printing Office, Washington, D.C., September 1995.

8

Managing the Networks

An amazing invention, but who would ever want to use one?

— Rutherford Hayes, U.S. President,
after participating in a trial telephone call between
Washington and Philadelphia in 1876.

Network Topography

The first telephone exchanges were manual, and the operators could connect any telephone to any other telephone connected to an exchange. The network for a single exchange office could be configured manually as a point-to-point, or "mesh," network, as in Figure 8-1. When automatic switches came into use, the central switching network for the exchange was reconfigured as a star, as shown in Figure 8-2. As connections between telephones in different exchanges and connections to long distance networks began to be needed, the telephone system grew into a hierarchical network where exchanges were interconnected by several levels of switches.

In the Days of "Ma Bell"

The PSTN was originally set up as a point-to-point system to connect a telephone at one location to another telephone at another location. The network was designed to provide dial-

Figure 8-1   Point-to-Point or Mesh Network.

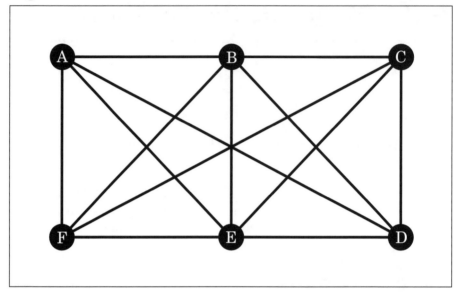

Figure 8-2   Centrally Switched Star Network.

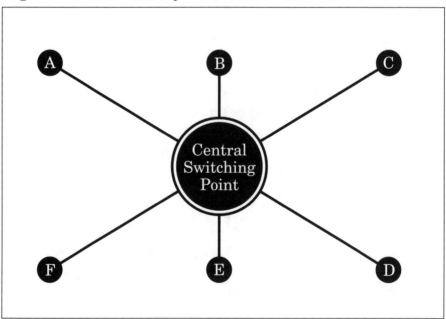

up analog channels with a frequency range of 300 to 3,000 Hz, between point A and point B. After a connection is established by this network, the channel is in use until the call is finished and the circuits used to establish it cannot be used for anything else.

Prior to deregulation of the telephone companies and the forced divestiture of the 23 Bell operating companies by AT&T, the network consisted of thousands of local telephone offices of the Bell system and independent telephone companies interconnected through the five-level switching hierarchy shown in Figure 8-3. Automatic switches, in response to control and supervisory signals, directed the progress of the calls, returned signals that reported the progress of the calls, and provided automatic accounting for toll calls.

Figure 8-3   North American Switching Hierarchy.

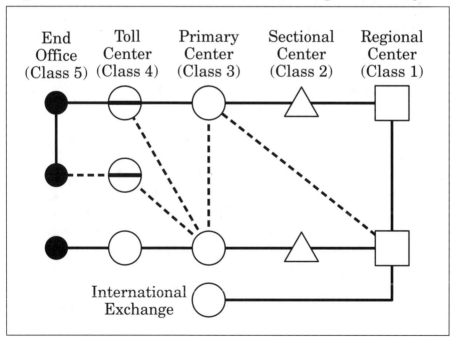

PSTN Hierarchy

The Bell system provided 85% of the local telephone service, GTE provided approximately 10%, and the balance was provided by the 1,400 independent telephone companies. Almost all networks followed Bell system practices.

Customers were connected to the lowest level of the hierarchy, the end office through a local loop. The connection between two telephones could be made through a single telephone office for local calls, or through multiple offices in different cities for long distance calls. The calls were routed through the lowest possible levels of the hierarchy. An office could be served directly by any higher ranking switch. For instance, a class 5 office could be served by a class 4 or higher level office. Similar hierarchies were installed by the telephone networks in Europe. Because the countries were smaller than the United States, and therefore had fewer telephones, the hierarchies did not need as many levels.

Interface between the North American hierarchies and international circuits was made through a gateway exchange in accordance with the recommendations of the Consultative Committee on International Telegraphy and Telephony (CCITT), now ITU-T.

The connections between the switching centers in the hierarchy were made through trunks (Figure 8-4). The transmission media used for these trunks varied from FDM analog carrier to coaxial cable, PCM carrier, and microwave radio, and, beginning in the late 1970s, fiber optic cable. An end office could be connected directly to an adjacent end office by a direct interoffice trunk. A tandem trunk connected an end office to a tandem switch in a class 4 office. Tandem switches are cross-connection switches that determine how to route and cross-connect a call to its destination over the shortest path

Figure 8-4   ITU-T Hierarchy.

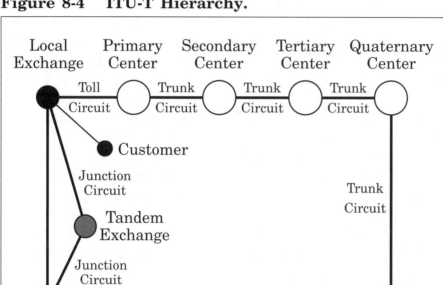

possible. Toll-connecting trunks connected an end office to a toll office. The dashed lines in Figure 8-3 indicate some of the many ways that offices could be interconnected when setting up a call. Alternate routing could be provided with this configuration in the event of heavy traffic or an outage of one or more of the trunks.

Signaling

The calls were set up by users dialing a rotary telephone dial, or by entering a number sequence on a keypad for the address of the called station. Initially, the network was managed and calls were controlled in response to signals within the voice channel. When the signaling consisted of tones within the voice frequency pass band of the channel, it was known as in-band signaling. When the tones were above or below the voice frequencies, it was known as out-of-band signaling (Figure 8-5). The format and specifications for these channel-associated signals conformed to a series of recommendations by the CCITT, beginning with CCITT System 1 and evolving to CCITT System 5. When forerunners of today's hackers developed tone-generating boxes that could duplicate tones and allow the theft of long distance service, CCITT Signaling System 6, a signaling system that used a separate channel for transmitting all of the signals for a group of trunks, was adopted internationally. This method of signaling is known as common channel signaling, and the AT&T Common Channel Interoffice System (CCIS) was installed extensively by the Bell system in the United States.

With the advent of digital switches, pulse code modulation (PCM) transmission and a digital multiplexing hierarchy, CCITT Signaling System Number 7 (SS7), for use in digital

Figure 8-5   In-Band and Out-of-Band Signaling.

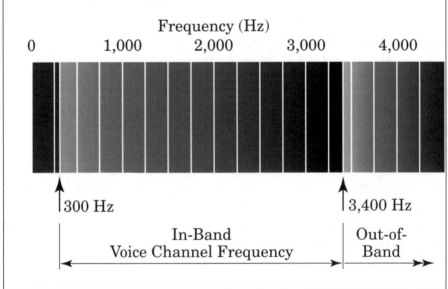

networks, was introduced in 1980. SS7 is a common channel system that was designed to utilize 64-Kbps digital channels of the transmission system for data links. In the United States, one of the 24 channels in a T1 multiplexer is dedicated to signaling. In any of the countries that use the Consortium of European Posts and Telegraphs (CEPT) standards, it can also use one of the time slots of the E1 multiplexer.

Divestiture and Competing Long Distance Carriers

When AT&T was required to divest itself of all of its operating companies in 1982, seven RBOCs were set up and the United States was divided into 160 local access and transport areas (LATAs). The RBOCs were granted the right to provide local telephone service transmission and switching within a LATA and were prohibited from providing service between LATAs. The interLATA trunks and the national toll network that had been part of the Bell system were taken over by AT&T, which became an interexchange carrier (IXC). Only the IXCs were allowed to provide service between LATAs. This provided opportunities for specialized common carriers such as MCI and Sprint, who had already built networks, to compete with AT&T for long distance service.

The operating telephone companies were required to provide equal access to any IXC, and the IXC was required to pay fees to the LEC for access to the local exchange network. Each IXC was allowed to have a point-of-presence (POP) in the class 5 offices of the LECs which usually consisted of an access tandem switch that provided the gateway between the LEC and the IXC (Figure 8-6). The subscribers were required to designate any one IXC as their primary long distance provider, and when the subscribers dialed a "1" for long distance access, they

Figure 8-6   IXC POP and Local Access.

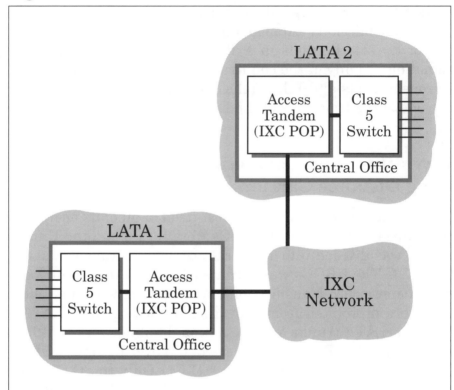

were connected to the designated IXC. Additionally, different five-digit addresses beginning with "1" and "0" (10XXX) were assigned to each IXC to provide an additional method for the subscriber to access any desired long distance provider.

This arrangement became effective January 1, 1984, and is still in place today, except that an additional "1" and "0" have been added to the access numbers so that 10-10XXX must be entered to access a particular provider. The operating companies own and control the infrastructure for local call (intraLATA) telephone service, and the interLATA circuits are provided by the IXCs.

Addresses and Numbering Plans

Every subscriber line in the world has a telephone number that provides a unique address for routing calls to that line through the switches of the network. This number directs the switches at the various nodes of the network.

Telephone numbers are assigned in accordance with numbering plans. We are familiar with the 10-digit format of the North American numbering plan, which divides the United States and Canada into geographic numbering plan areas (NPAs). Each NPA is identified by a unique number called an area code. Only seven digits are required to place a call within any NPA. The first three digits select an exchange within the NPA, and the final four digits select a particular line within that exchange.

When placing a call outside the calling NPA, the digit "1" must first be dialed to connect the calling line to the POP of a previously designated interexchange carrier. The three digits following the "1" select an area code, the second three digits select an exchange, and the final four digits select a subscriber line.

The digits can be dialed by pushing numbered keypad buttons on a dual-tone multifrequency (DTMF) telephone or by using a rotary dial. The DTMF telephone sends different combinations of dual tones to the central office switch for each of the buttons on the keypad. If the button for number 9 is pressed, the telephone sends 852-Hz and 1,477-Hz tones simultaneously. The rotary dial telephone sends from 1 to 10 timed open/close pulses over the line for each digit dialed. For example, if the number 4 is dialed, the loop connection will be opened and closed four times.

International Calling

International calls may be dialed directly from DTMF-equipped telephones. The ITU-T recommends a limit of 12 digits for international telephone numbers. In the United States, international calls may be originated by dialing "011," the international access code.

International Access Code	Country Code	City Code	Local Telephone Number
011	33	556	321 4567

The Long Distance Carrier Networks

The plant for both local and long distance service has been enormously expanded and modernized with new technology since divestiture and the creation of the IXCs. The three major long distance carriers, AT&T, MCI, and Sprint, have spent billions expanding and upgrading their networks to meet the demand for data transmission, which has been increasing at a rate of 85% per year.

Sprint

Initially, the Sprint network utilized microwave radio transmission, and customer access was through the LEC network by means of access tandem switches at the Sprint POP. In 1984, when deregulation of U.S. telephone service and divestiture of the Bell operating companies by AT&T became a reality, Sprint announced plans for overlaying its system with a completely digital fiber optic transmission network. Initially, Sprint had overlaid legacy T1 and T3 circuits with a fiber optic network that operated at a transmission rate of 622 Mbps. The new fiber optic network, which was built in rings across the country, was completed in 1987. By 1988, Sprint had also deployed SS7 throughout its entire network.

Sprint began nationwide deployment of SONET circuits over fiber in 1994. The fiber was deployed in a four-fiber, bidirectional, line-switched ring (4-FBLSR) configuration. By the end of 1997, 100 rings had been built, with 170 expected to be completed by the end of 1998. By the middle of 1999, all of Sprint's traffic, voice, data, and video will be transported over the SONET ring infrastructure.

The new Sprint network (Figure 8-7) is based on a three-tier architecture: an ATM backbone core, ATM edge switches (located at 29 nodes on the periphery of the network), and the broadband metropolitan fiber optic/SONET ring networks (BMAN) that extend Sprint's reach to 70% of large businesses in the United States. For locations that do not have access to BMANs, universal ADSL lines are used to provide access. Edge switches can also be installed on customer premises.

The ATM backbone core network consists of eight high-speed ATM switches, with throughput capacities from 10 Gbps to 160 Gbps, interconnected by a series of 4-FBLSR rings (Figure 8-8). The 29 edge switches are connected to the core network by additional four-fiber SONET rings.

The bidirectional, line-switched rings that provide transmission and protection for the system have two working fibers and two standby, or protection, fibers. The two working fibers transmit and receive traffic around the ring in one direction, for instance from A to B to C in Figure 8-9. In the event of a fiber cut or an electronics failure, the traffic can be switched to the two standby fibers in approximately 200 ms (depending on the length of the path) and transported around the ring in the

Figure 8-7   Sprint Network. (Courtesy of Sprint.)

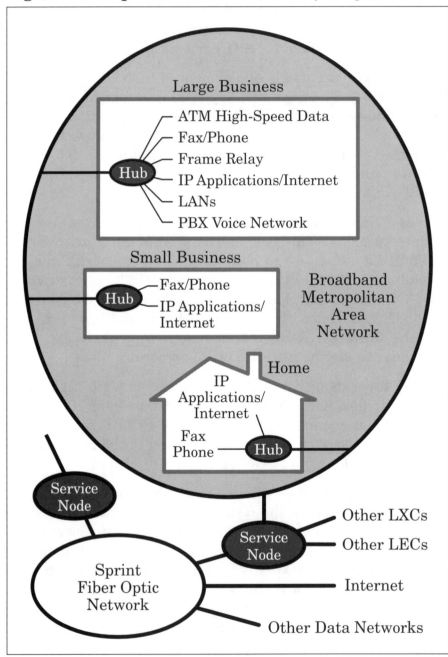

Figure 8-8   Rings and Nodes of Sprint Core ATM Network. (Courtesy of Sprint.)

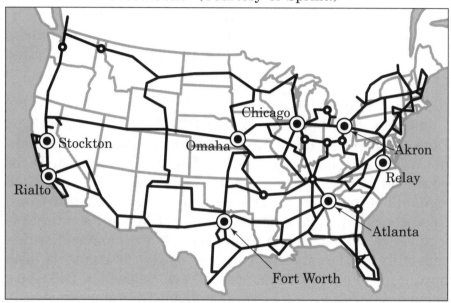

Figure 8-9   Bidirectional Line-Switched Ring.

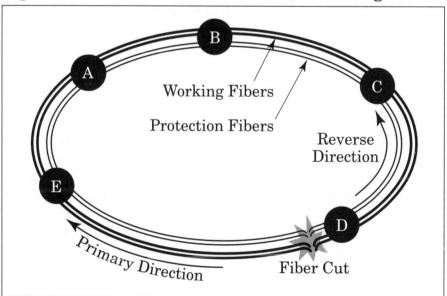

reverse direction. For instance, in the event of a cut between nodes D and E, traffic from A, B, C, and D would be transmitted to node E over the protection fibers in the reverse direction. In addition to the protection provided by switching fibers in the rings, the network can also provide protection against outages by switching to alternate paths.

Initially, the transmission over each fiber in the rings was a single wavelength of light at the SONET OC-48 rate of 2.5 Gbps. In 1996, Sprint began equipping the rings with 16-channel WDM, so that 16 wavelengths of light are transmitted over each fiber. Each light wavelength is modulated with an OC-48 signal. This has increased the capacity of each ring to 40 Gbps. Recently, Sprint began installing new WDM equipment that is initially capable of transmitting 40 lightwave channels, with a capability for the capacity to be increased to 96 channels. The maximum transmission capacity of each ring will be 240 Gbps, enough capacity to carry 3 million phone calls simultaneously over a single fiber pair.

The interconnection of rings in the Sprint three-tier network is illustrated in Figure 8-10.

MCI WorldCom

MCI originated in 1963 as Microwave Communications, Inc. with a microwave link between St. Louis and Chicago. As its original name implies, its national network was at first tied together with microwave radio links. Today, the microwave network has been replaced with over 40,000 route miles of optical fiber cable. The network is 100% digital, operating at digital data rates from 2.5 Gbps to 80 Gbps. The backbone network uses 8-wave WDM to combine OC-192 (10 Gbps) digital channels to provide 80-Gbps transmission over each fiber.

Figure 8-10   Sprint Network: Rings Interconnection.
(Courtesy of Sprint.)

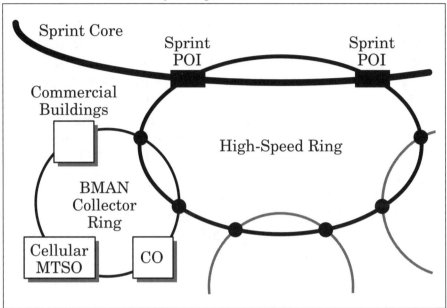

Nearly all of the current network is fiber optic cable, with connections to some remote locations made by digital microwave
radio. The network combines SONET and ATM technologies.

Over 70% of MCI's long distance traffic is carried over SONET
network circuits. By mid-1998, MCI had deployed 38 bidirectional SONET rings around metropolitan areas. Additionally,
78 local SONET networks have been built in 38 major U.S.
cities. In the event of a cable cut, the SONET rings can restore
service in 50 ms.

MCI's worldwide network is managed by network MCI Services.
At the global network management center in Cary, North Caro-
lina, management personnel utilize computerized monitoring
systems to monitor the physical portion of the network 24 hours
a day, 7 days a week. Network data are constantly analyzed to

recognize and correct potential problems before traffic is affected.

The MCI network provides many alternative pathways for routing telecommunications traffic. In addition to the fiber optic rings, the domestic network contains approximately 400 digital cross-connects (DXCs), which can remotely reroute traffic around a damaged path in the event of an outage. The network uses real-time restoration (RTR) to orchestrate the restoration. RTR develops restoration plans, implements pre-plans, and sends routing messages to the digital switching centers (DSCs) to establish a transmission path around the outage. The network uses another tool, dynamically controlled routing (DCR), to dynamically define many routing paths every few seconds, based on traffic conditions, to find routes around any problem.

In August 1997, MCI began the field trial of an optical cross-connect system as the first step in the creation of an all-optical network. An all-optical network eliminates network dependency on electronics, increasing network efficiency and increasing capability to handle higher transmission rates. In an optical network with electronic cross-connects, information must be converted from light to an electronic signal before it is re-routed and converted back to a light signal. In an optical cross-connect, different wavelengths of a WDM optical signal on a fiber, each modulated with data rates as high as 10 Gbps, are demultiplexed by filters and optically switched to other fibers that route them to desired destinations. High-density circuits can be switched in microseconds to restore circuits and restore outages.

AT&T

At the time of separation from Ma Bell in 1984, the extensive AT&T network provided 97% of the long distance service in the United States. The network conformed to the North American switching hierarchy of Figure 8-3. It was so extensive (present in almost every central office in the United States), that in the event of an outage on one path, service could be restored by alternate routing. Much of the network transmission was over microwave circuits, with some fiber optic circuits in high-density areas such as the northeast corridor.

Today, the AT&T network has been overlaid with 40,000 route miles of optical fiber cable, which transports more than 97% of AT&T traffic within the United States. The network is 100% digital for all switched traffic, and voice circuits are multiplexed at SONET rates. Over 1,000 route miles of fiber optic cable have been equipped with 8-wavelength WDM to date, increasing capacity to a rate of 20 Gbps for each fiber pair on these routes. Sixteen-wavelength WDM is being tested, and 80 wavelength WDM is expected to be installed in the near future. SS7 is deployed throughout the network for transporting the signaling information that directs call setup and management of call progress through the network.

The network features layers of protection against route outages. Real-time network routing (RTNR) routes calls over any of 134 possible routes and reroutes traffic in the event of an outage. FASTAR® and FASTAR II® automatically reroute entire circuits when a fiber optic cable fails. Service is restored on most circuits within 2 to 3 minutes after a cable cut.

AT&T is currently overlaying the network with 58 regional bidirectional, switched rings. Transmission over the rings will be SONET based at the OC-48 rate of 2.488 Gbps. The rings will utilize WDM to increase the information transport over each fiber pair to as high as 200 Gbps (if 80 wavelengths are implemented). As described in the discussion of switched rings above and illustrated in Figure 8-9, protection is provided by switching from working fibers that transmit and receive traffic around the ring in one direction, to standby fibers that transport the traffic around the ring in the opposite direction. AT&T expects to be able to restore service in 50 ms.

New Age Broadband Carriers

A number of new transcontinental backbone networks have emerged in the United States in the past 3 to 4 years. Companies, such as railroad and pipeline operators, that own right-of-way have installed fiber optic cable and electronics along them. Currently each of these networks have deployed from 4,000 to over 10,000 route miles of fiber cable, with each network's projected deployment by mid-1999 to be 13,000 to 18,000 route miles.

These networks employ the very latest technology, including ATM transmission over bidirectonal, line-switched SONET rings, WDM, and the most advanced commercially available fiber. Some of the networks have high fiber counts in their cables and employ WDM with as many as 16 wavelengths, each modulated at data rates up to 10 Gbps to achieve a capacity of 160 Mbps on each fiber and a total capacity of 23 Tbps on their network. All of the networks employ SS7 in network control. The salient technical details of five of the new networks are presented in Table 8-1.

Some of these carriers, known as carrier's carriers, offer network capacity to local telephone companies, regional carriers, broadband service providers, Internet service providers, long distance carriers, and cable television companies. Some of the

Table 8-1   **New Age Fiber Optic Networks.**

Carrier	Projected Route Miles (1999)	Cities	Core Network WDM Cross Section (Fibers)	Channels per Fiber	Optical Rate per Channel	Total Capacity (Gbps)	BSL Rings
Frontier Corp.	15,000	120	24	16	OC-48	3,840	Yes
IXC Comm., Inc.	20,000	70	22	8	OC-48/ OC-192	880	Yes
Level 3 Comm., Inc.	15,000	50	96	N/S	N/S	N/S	N/A
Qwest Comm., Corp.	18,449	130	48	N/S	OC-192	480	Yes
The Williams Co., Inc.	32,000	69	96/144	16	OC-192	15,000/ 23,000	Yes

N/A – Information not available   N/S – Not specified

companies offer private-line SONET-based OC-3 and OC-12 leased circuits. Some of them offer dark fibers for lease. Some are extending their networks by swapping fibers with other carriers. Some of the networks plan to build local networks in cities across the country and interconnect them with a national long distance network. Most of them have connections with international networks. One of the networks claims to be the first international network optimized for Internet technology because it uses the Transmission Control Protocol/Internet Protocol (TCP/IP) from end-to-end.

Press releases for the networks boast of computerized monitoring systems and at least 99.98% availability through nearly instantaneous restoration of outages by self-healing rings.

Cellular Radio Networks

Cellular radio communications is a subset of wireless communications in which the geographical areas are divided into cells, with a transceiver and antenna located within each cell. Each of the transceivers can communicate with a cellular telephone, (which also contains a transmitter, receiver, and antenna), over a radio frequency channel. There are several hundred radio frequencies available for use within a geographic area, so that nearby cells may be assigned different frequencies. The power radiated by the transmitters within each cell is low, allowing reuse of frequencies by geographically separated cells.

Each cellular telephone is assigned a unique number. In the United States and Canada, 10 digit numbers are assigned in accordance with the North American Numbering Plan (NANP).

The cellular telephone network must be able to provide customer mobility by handing off a cellular telephone as it moves from one cell to another while a call is in progress, even if the mobile switching centers for the cells involved belong to different service providers. Intrasystem handoff as the mobile travels between cells of the same system is controlled by proprietary methods at the mobile switching centers.

Functions that enable customer mobility in a mobile telecommunications network include: radio system management; mobility management; call processing; service management operations, administration, and maintenance (OA&M); and terrestrial transmission facilities management. Seamless roaming and intersystem handoff are subsets of mobility management.

Radio System Management

The development of diverse analog standards in the different countries of Europe resulted in a "tower of Babel" that required a different mobile telephone for each country entered. This prompted the development of a single digital continental compatibility standard based on time division multiple access

(TDMA) and that includes signaling specifications as well as radio transmission specifications. A new frequency spectrum in the 900-MHz region was assigned for the new service, which has been named the Global System for Mobile Communications (GSM). Many of the countries in South America have also adopted the GSM standard.

In the United States the first-generation technology used in cellular radio was frequency-modulated (FM) analog radio and was named the Advanced Mobile Phone System (AMPS). Subsequent generations, narrowband AMPS (NAMPS), are also FM analog. Three new transmission technologies have been introduced and specified in standards. In the absence of a national standard in the United States, each cellular company is free to use whichever air interface standard it chooses. The emerging digital technologies, TDMA and code division multiple access (CDMA), are being deployed by many cellular providers, requiring mobile telephones to be more complex and expensive to provide roaming between systems with different technologies.

Proposals for a third generation standard to be called IMT 2000 are under review by the ITU-T.

In Japan there are four cellular systems in use, two analog and two digital TDMA systems.

Mobility Management

In the United States mobility management is provided by the protocols of the ANSI standard ANSI/TIA /EIA-41, also referred to as IS-41. This standard defines intersystem handoff to maintain automatic seamless roaming when the mobile travels between systems or originates or responds to calls while away from its home system.

In intersystem operations IS-41 is credited with supporting the following three basic mobility functions: intersystem handoff, automatic roaming, and intersystem OA&M. Processes are specified by IS-41 for handoff, automatic roaming, authentication, call processing, and OA&M functions. Included are methods and protocols for providing subscriber identification, mobile station identification, electronic serial numbers, authentication, and billing identification.

Call processing involves all network signaling, management, and connectivity required to establish, maintain, and release calls between a mobile subscriber and a wire line phone or another cellular phone. The phones involved may be served by different networks, and the calls may be mobile originated (calls that are placed from a mobile station) or mobile terminated (calls that are made to a mobile station). IS-41 uses SS7 for the transfer of signaling information through the mobile communications network.

Cable TV Networks

The management of early community antenna television (CATV) networks consisted mainly of maintenance of the network to ensure reliable service to subscribers and expansion of the network to meet new subscriber requirements and increase revenue.

The CATV multisystem operators (MSOs) are overbuilding their networks today to provide megabit digital transmission in the upstream and downstream directions to provide many new services. The new services planned include voice and data with Internet access at megabit rates, long distance telephone, digital high-definition TV, interactive TV, and video-on-demand. In addition to new services for subscribers, the cable TV networks have begun to lease bandwidth to PCS providers for distribution of mobile telephone calls to microcell transmitter/ receiver sites.

Providing these services requires more complex broadband cable networks and more sophisticated network management. Most MSOs have overlaid their original coaxial cable trunk systems with fiber optic systems that include WDM and fiber amplifiers and provide transmission in both the forward and reverse directions.

Many cable TV companies have adopted the H-F/C technology, using the bandwidth and frequency allocation shown in Figure 7-12. A generic H-F/C CATV network shown in Figure 8-11, utilizes primary and secondary bidirectional switched fiber rings feeding a tree system. The secondary fiber rings are connected to the primary fiber rings through optical hubs. Optical nodes on the secondary rings feed optical/electronic transition nodes in a star configuration. Coaxial cable trees branch out of the transition nodes to the premises served. Forward and reverse RF amplifiers in the coaxial cable branches provide two-way transmission. A network interface modem or a set-top box provide separate output ports for each of the services. Many other architectures for CATV networks, such as a ring-in-ring, mini-star, and dual-ring star/bus, have been installed or are planned. Most use an H-F/C approach to reach the subscribers.

Managing the CATV networks so that they can provide reliable voice and data service is a key to success in the local and long distance telephone business. The customers have come to expect 99.975% availability of the telephone system. To achieve the required reliability, the MSOs are installing sophisticated operations support systems and are planning to convert network powering systems to central node power. The operations support systems include network surveillance

Figure 8-11   CATV System.

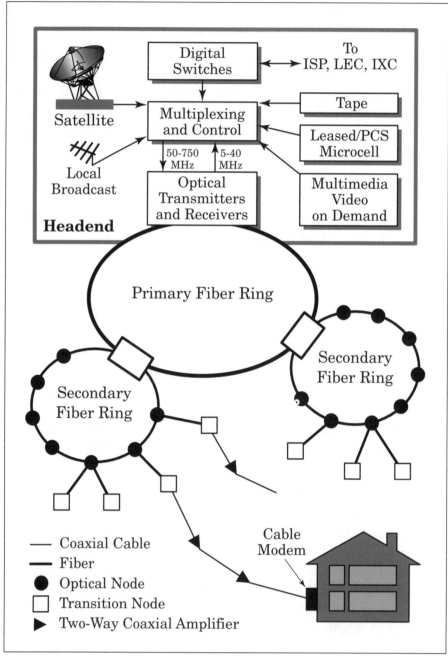

systems to monitor the status and performance of the network elements in real time, and to automatically reconfigure the nework to prevent impending service outages before customer service is interrupted.

Supplying central node power to the network means that the network, the cable modems at the customer premise, and the telephones inside the premise will be powered from an independent network source instead of depending on a power utility's grid. In the event of an emergency that causes loss of the utility's power, the telephone will still be operational. This is extremely important to customers, since in an emergency, life or death may depend on working telephone service.

Managing the Internet

The Internet is a vast group of worldwide information sources which can be accessed by over 30 million computers through several thousand computer networks interconnected in real time by means of TCP/IP protocols.

The Internet is a packet switched network. Data streams are broken up into packets, which are moved through the network in accordance with rules, called protocols. TCP/IP was developed under the sponsorship of ARPA, which formally adopted it in 1983. The TCP protocol determines how the data to be forwarded are broken into packets. The packets are numbered and a section called a header, which contains the packet identification number and other information about the packet, is added. The data packets are assembled in larger IP packets, which contain the addresses that tell the packet switches where to send the packets. Each router along the route reads the address and forwards the packet to another router. There are many routes between the originating and terminating computers, so different packets may not take the same route to the destination and will not necessarily arrive in order. At the destination, the packets are reassembled according to their ID number. Because there are so many paths over which the packet may be sent to a particular destination, the packets can still reach their destination despite an outage in one of the pathways.

No one manages or governs the Internet directly. There are, however, several volunteer organizations that solve technical and operational problems, develop and approve standards, and ensure compatibility of hardware and software from different vendors.

The Internet Society (ISOC) is an organization of volunteers that contributes the primary oversight for the direction of the Internet. It appoints the members of the Internet Architecture Board (IAB), which authorizes and adopts standards and determines how Internet addresses are assigned.

The Internet Engineering Task Force (IETF) is another organizaton of volunteers that forms working groups to consider and solve technical problems and to make recommendations and develop standards that are presented to the IAB for approval and adoption. The IETF working groups are grouped in areas, managed by members of the Internet Engineering Steering Group (IESG), who are called area directors.

Conclusion

Transmission technology and network management tools have advanced to the point where the backbone networks can offer better than 99.9% availability. In order to accommodate the increasing data transmission capacity throughout the world increasing at an exponential rate, it is necessary to overbuild of existing networks and install new networks.

Bibliography

AT&T: Discussions with AT&T personnel and material from the *AT&T Fact Book*, available at www.att.com/factbook

Bell Telephone Laboratories, *Transmission Systems for Telecommunications,* Winston Salem, Western Electric Company, NC: 1970.

Blyth, W. John and Blyth, Mary M., *Telecommunications: Concepts, Development, and Management,* Mission Hills, CA: Glencoe Publishing Company, 1985.

Dern, Daniel P., *The Internet Guide for New Users,* New York, NY: Mc Graw-Hill Inc., 1994.

Freeman, Roger L, *Telecommunication System Engineering,* New York, NY: John Wiley & Sons, 1980.

Hahn, Harley and Stout, Rick, *The Internet Complete Reference,* Berkeley, CA: Osborne McGraw-Hill, 1994.

IETF Secretariat, "Overview of the IETF," available at www.ietf.org/overview.html

Krol, Ed, *The Whole Internet User's Guide and Catalog,* Sebastopol, CA: O'Reilly & Associates, Inc., 1992.

Martin, James, *Telecommunications and the Computer,* 2nd ed., Englewood Cliffs, NJ: Prentice-Hall, Inc, 1976.

Muller, Nathan J., *The Totally Wired Web Toolkit,* New York, NY: Mc Graw-Hill, 1997.

Paulson, Ed, *The Complete Communications Handbook,* Plano TX: Wordware Publishing, Inc., 1992.

Pearce, J. Gordon, *Telecommunication Switching,* New York, NY: Plenum Press, 1981.

MCI: Numerous press releases and material. Information available at www.mci.com/aboutus

Sprint: Numerous press releases and diagrams. Information available at www.sprint.com

9

The Digital Wireless World

Voice communication over wire, although technically possible, hardly seems a reliable means of conversing, and if it were, it is unlikely that the public would avail themselves of such a service.

— Boston Herald, 1891.

Wireless communication services are expected to be the telecommunications industry's most significant growth area in the next decade. However, some industry analysts believe that this technology is more hype than reality. But then again, that was the same response that many people had to the Internet.

Cellular Radio

Cellular radio has proven to be one of the fastest-growing technologies in the world. The first system was tested in Chicago in 1977 and placed into full commercial service in 1983. In order to stimulate competition, the FCC decided that two licenses would be issued in each area. Metropolitan areas are the best markets for the present terrestrial systems. The systems are designed to grow gradually to serve more users as demand increases. The larger markets have led to lower prices for customers.

The market for wireless equipment, such as cordless phones, pagers, and cellular phones, has continued to expand. At the end of 1991, there were 7.6 million people in the United States using cellular phones. The number of cellular users increased to over 55 million by 1997.

Personal Communications Services (PCS)

The introduction of PCS using pocket-size digital cordless telephones has opened up a huge market. This has created direct competition between local telephone companies and cable TV companies for a share of this market. This development will enable people to be reached anywhere in the wire line, wireless, and cellular networks. Subscribers will have a personal identification (ID) number, instead of a location-dependent telephone number. This ID number will enable callers to reach subscribers at a single number, regardless of whether they are at home, in the office, or traveling. To paraphrase Arthur C. Clarke, the time is coming when you will call someone, and if they do not answer, you will know they must be dead.

Handsets for PCS are similiar to traditional cellular telephones on the outside. But because PCS handles phone calls as a stream of digital bits, like the way a computer handles data, the system offers built-in features such as numeric paging, easier data transmission, and increased security. Calls can't be overheard with a scanner, and the phone number and other transmitted data are encrypted. Handset manufacturers include Nortel Networks, Motorola, Ericsson, and Nokia. The handsets are getting smaller and lighter. The Nokia 6160 weighs less than 5 oz (Figure 9-1). It has up to a 14-day standby and up to 5-hour talktime before the battery needs recharging.

The demand for PCS will be on a global basis, but the regula-

Figure 9-1   Nokia 6160 Handset.

tions that govern the allocation of radio spectrum are determined country by country.

In the United States and Canada, the 2-GHz band has been allocated for PCS service. In the year 2000, the World Radio Conference (WRC) will identify additional spectrum for the third generation standard, known as IMT 2000 or the "millennium phone."

In the United States, the FCC auctioned off the 2-GHz spectrum to 22 bidders for 99 licenses, bringing in a grand total of $7.7 billion. However, the spectrum was held by other licensees. The PCS providers had to negotiate the surrender of the spectrum by providing the incumbent with new facilities that were technically equivalent to those being replaced. In addition, the PCS industry has spend another $10 to $20 billion on new infrastructure.

The Standards

There are three standards in operation worldwide: GSM, TDMA, and CDMA. Manufacturers for each technology have made sure that they have products that will support both PCS and digital cellular hardware, which are basically similar technologies.

GSM dominates the digital cellular marketplace outside of North America. It has been adopted in 86 countries by 156 operating companies and was first commercially available in Europe in 1992, where it is now a relatively mature product. It is the primary cellular and PCS service for the majority of the digital wireless voice subscribers worldwide.

TDMA is a derivative of GSM. It is a market contender chiefly because it is being supported and promoted by AT&T, which, because of its $12.6 billion acquisition of McCaw Cellular, is now the largest analog cellular carrier in the United States. Like GSM, to which it is closely related, TDMA allocates bandwidth to subscribers by splitting it into time segments.

CDMA is a new technology that has many advocates in the United States. CDMA spreads the signal over multiple frequencies and reintegrates the transmitted intelligence using code assignments. In this regard it has its roots in decades-old spread-spectrum technology first used by the U.S. military. The advocates of CDMA point toward its more efficient use of the spectrum and excellent signal clarity under congestion as reasons for its adoption.

Service providers that are committed to CDMA include Sprint, GTE, Ameritech, AirTouch, Bell Atlantic, Nynex, and US West. The seven companies licensed to develop GSM-based PCS networks in the United States are American Personal Communications, American Portable Telecom, Bell South Personal Communications, Intercel, Omnipoint, Pacific Bell Mobile Services, and Western Wireless Co. Together this group holds licenses covering over 125 million Americans, including 12 of the largest 25 cities.

Two new standards have been proposed to the ITU for a third generation of cellular systems. Both proposals are based on CDM, and are similar, but they are not interoperable, so global roaming will not be possible unless a compromise is reached. A proposal that manufacturers produce a multimode handset that will operate with either system has also been made to the ITU, which is currently considering arguments for and against convergence of the two CDMA standards.

The Canadian Market

The two incumbent analog cellular providers, the Mobility Canada Consortium and Rogers Cantel Inc., have built nationwide PCS networks. The two new PCS providers with nationwide networks are Microcell Telecommunications Inc. and Clearnet Communications Inc. Both companies have been enjoying strong subscriber growth, but they are not expected to make a profit in the near future.

Some experts believe that the copper wire loop that carries phone service to the home is destined to become history. Bell Canada, which has a huge investment in copper wire, is moving to wireless. Rural customers around Chatham, in southwestern Ontario, recently became the first in North America to have their phone service linked to the network by radio signals rather than conventional copper cables.

A move toward a single number has also been made by Bell's Simply One program. Customers use a single number for their phone, voice mail, and cellphone. If a call coming to your home isn't answered, it automatically rings your cellphone. If that isn't answered, it is switched to a voice mailbox.

The design goal is to produce something like a pocket-sized organizer that has the capability to provide all data services, including Internet, video, and multimedia. An example of how this might work is already available. A Nokia 9000 Communicator is a digital phone with a text keyboard that contains an equivalent to a 386 desktop computer. It can send faxes and e-mail and has the ability to link to the Internet through the digital phone system.

Global PCS

The concept of a global PCS utilizing a network of satellites extends the reach of terrestrial PCS and will make personal communication networks virtually ubiquitous.

Until now, international telecommunications satellites have operated from an orbit some 36,000 km above the Earth. From this geostationary Earth orbit (GEO), the satellites appear to remain fixed above a single spot on Earth.

The coming generation of personal satellite communications systems would place satellites in orbits closer to Earth to provide hand-held mobile phone service.

The global PCS network, based on low Earth orbiting (LEO) satellites at altitudes of about 1,000 km, and medium Earth orbiting (MEO) satellites at altitudes of about 10,000 km, and forming a chain around the globe, is anticipated to begin commercial service in 1998 and 1999 (Table 9-1).

Table 9-1   Global PCS Network.

Network	Type	No. of Satellites	Altitude (km)
Iridium	LEO	66	780
Globalstar	LEO	48	1,400
Ellipso	LEO	17	520/8,040
ICO Global	MEO	10	10,355

These satellite systems include Iridium, backed by Motorola; Globalstar, backed by Loral Corp. and Qualcomm; ICO Global, a spin-off of the international treaty organization Inmarsat; and Ellipso, backed by Lockheed Martin and Harris Corp. Geostationary satellites have a major advantage because continuous global coverage can be provided with only three satellites. However, the drawbacks include the round trip delay and high power required. The LEOs minimize the transmit power and the delay but require a large number of satellites to provide global coverage. In addition, the speed of the LEOs relative to the Earth (about 7.4 km/s) requires the calls to be switched more frequently from one satellite to another, a complicated and expensive technical challenge. Furthermore, because their satellites orbit so close to Earth, LEO systems are more susceptible to "shadowing," the blockage of signals by buildings or hills that sometime disrupts cellular service.

Iridium and Globalstar plan to use LEOs at altitudes of 780 and 1,400 km, respectively. ICO plans to use MEOs at an altitude of about 10,355 km. Ellipso, while technically a LEO system, will operate up to MEO altitudes using 17 satellites in three orbital planes. To provide full global coverage, LEO systems require scores of satellites. Globalstar, for example, will use 48 satellites (Figure 9-2), while Iridium plans on 66 satellites. The Iridium system will have six orbital planes, each with 11 satellites, as shown in Figure 9-3. The satellites will be phased so that odd-numbered planes have satellites in corresponding locations, and satellites in the even-numbered planes are staggered approximately midway between. The satellites will travel in co-rotating planes, up one side of the Earth, cross over at the pole and come down the other side of the Earth.

Figure 9-2   The Globalstar Network.

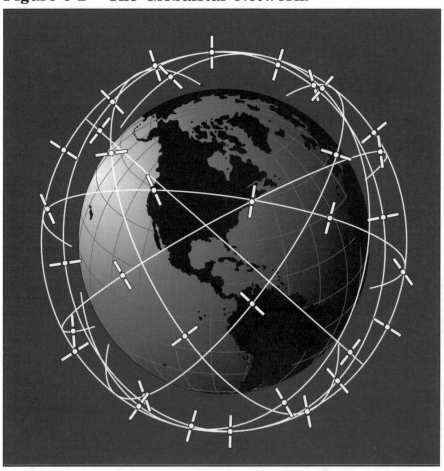

Figure 9-3   The Iridium Constellation.

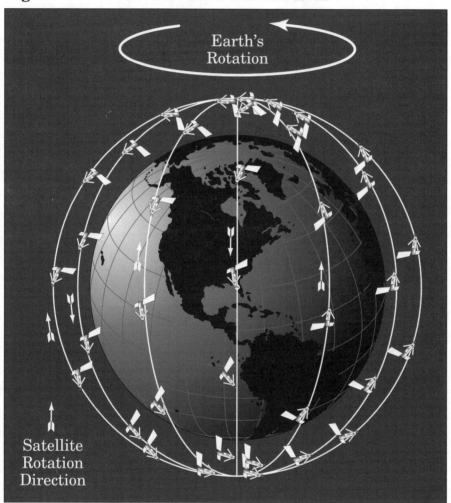

The Earth, of course, continues to rotate beneath them. As shown in Figure 9-4, when the satellite receives the signal, routing will be handled on Earth via a gateway station. The call can then be transferred from satellite to satellite and passed down to Earth. Communication link between the subscriber unit and the satellite will use the L-band (1,500 – 1,700 MHz)

Figure 9-4   The Iridium Network.

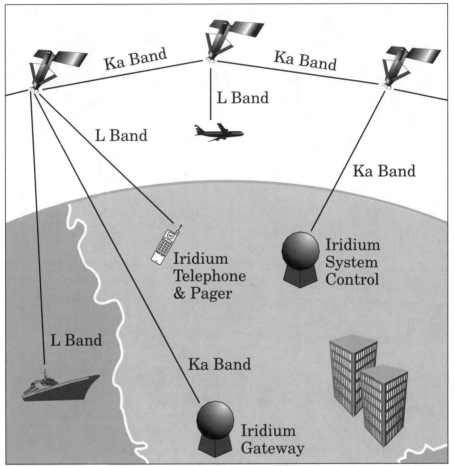

while intersatellite links and gateways will use the Ka band (17/30 GHz). A typical Iridium satellite is shown in Figure 9-5. In order to provide global service, LEO systems also require either complex satellite crosslinks, such as what Iridium uses, or large numbers of ground stations (Globalstar is planning 100 to 200 sites). Moreover, LEO satellites have to be replaced more often than MEO and GEO satellites (a design life of 7.5 years for Globalstar versus 12 years for ICO's MEO satellites).

Figure 9-5   Typical Iridium Satellite.

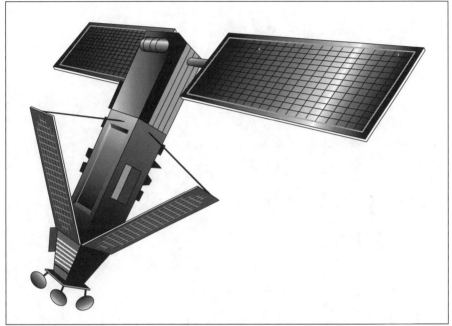

The MEO system provides several advantages in cost and reliability over LEO and GEO systems. A MEO system can cover every inhabited area on Earth with a small number of satellites and ground stations.

Satellites in the MEO orbit will also be "in sight" of a phone user for up to 90 minutes, dramatically reducing the need for switching calls from satellite to satellite. The GEO, MEO, and LEO orbits are shown in Figure 9-6.

The handsets for the global PCS network will essentially be palm-sized Earth stations. They will operate in both cellular and satellite modes. Where terrestrial service exists, the handset will operate normally. Where it is absent or interrupted, the handset will link directly and transparently to a satellite. Iridium's handsets and pagers are shown in Figure 9-7.

Figure 9-6   GEO, MEO, and LEO Orbits.

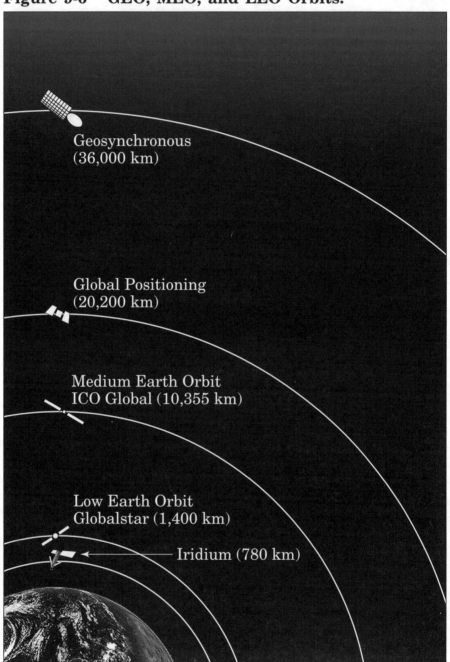

Figure 9-7   Iridium Handsets and Pagers.

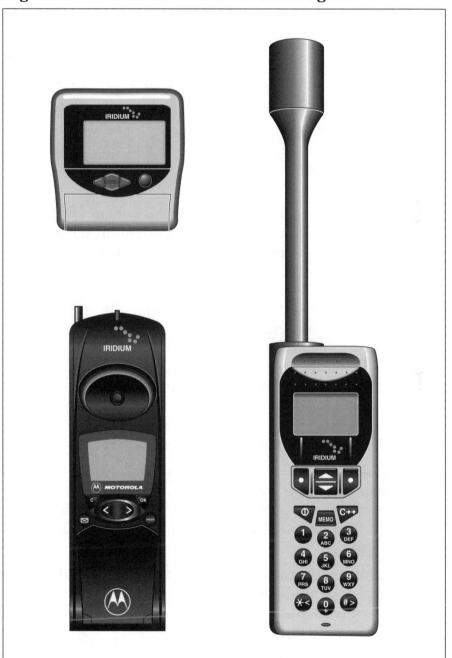

Internet Access

With the growth of the World Wide Web, operators are looking for faster, more efficient ways to transmit large amounts of digital data. While fiber to the home is the ultimate solution, it is still years away. Instead, satellite-based data networks can offer fast transmission speeds and global access.

Hughes Communications Inc. is building Spaceway, a $4 billion 20-satellite system equipped with spot beams that can provide access to the Internet and bypass overloaded links on the ground.

Craig McCaw, who sold his cellular phone system to AT&T for $12.6 billion in 1994, and Bill Gates of Microsoft have formed Teledesk Corp. They plan to use 288 small satellites to provide low-cost high-speed Internet access, corporate networking, and desktop video. The satellites will zoom across the sky just 1,400 km above the Earth. Each satellite will hand off Internet traffic to the next, as it moves over land and sea. It will require extremely sophisticated software and networking to link the satellites, not to mention the logistical challenge of getting that many satellites off the ground.

Wireless Local Area Networks (LANs)

Wireless LANs may be divided into two primary types, according to the nature of the transmission: infrared LANs and radio frequency (RF) LANs. In addition, linkages via laser transmission in free space are under development for building-to-building linkages.

Infrared LANs were the first to enter the market, but limited range, slow speed, and inability of systems to transmit through walls have limited the applicability of the infrared approach in diverse business and industrial settings.

Wireless RF LANs have proven the most successful. Depite the ubiquitous use of highly robust spread-spectrum transmission techniques in this type of equipment, careful installation and deployment remain absolute requirements.

A typical wireless LAN based on standard 18-GHz microwave technology is shown in Figure 9-8. The system is designed to extend or replace existing hardwired networks such as Ethernet or token ring. The system is capable of initially delivering 15 Mbps to the desktop and replacing the last 30m of wiring typically found in today's LAN installations. A ceiling-mounted transceiver interconnects up to 32 computers and printers in a cell about 25m in diameter. The system eliminates the cost of wiring an office and, more importantly, the cost of changing

Figure 9-8   Wireless LANs.

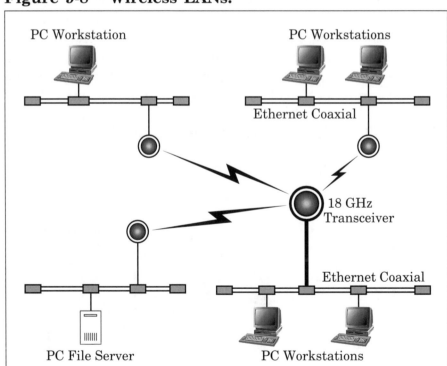

the wiring when an office is reconfigured or moved. Many current systems permit roaming, some to the extent that any node can operate anywhere as long as it remains within the range of another node. This capability, however, will depend upon the networking software used in the system at large. Some systems are capable of transmitting over many miles, thus connecting widely separated facilities.

Wireless LANs, for the most part, reflect older networking models such as Ethernet or token ring rather than the new intranet concept. When the LAN is totally wireless and dispenses with dedicated phone lines, the cost disadvantages of these older models disappear.

Connectivity topologies can be categorized into three general network classes: point-to-point, point-to-multipoint, and mesh (Figure 9-9). The point-to-point topology represents duplex communications between two users. In the point-to-multipoint topology, there is a single sender and multiple recipients. Duplex communications are possible only between the single sender and any one of the recipients. Mesh connections allow duplex communications between any node in the network. Of

Figure 9-9   Wireless LAN Connectivity Topologies.

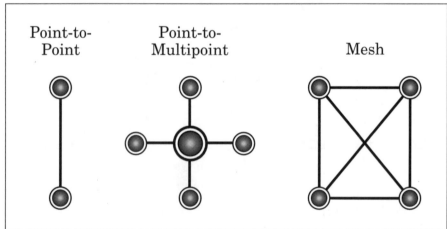

interest here are point-to-point connections, specifically those connections that can support high data rate duplex communications between fixed (such as building-to-building) or semifixed (building-to-mobile van) locations.

Wireless LAN equipment is employed in highly diverse applications today, but a favorite use remains the warehouse setting where employees with handheld terminals can move freely amidst the inventory, making notations as needed. A growing use is the sort of manufacturing operation where work spaces must frequently be reconfigured and terminals moved.

Conclusion

Whatever the future holds for wireless systems like digital cellular phones, PCS, Internet access, pagers, and LANs, there is no question that they have the potential to radically transform the telecommunications industry.

Global PCS operators plan to use their systems to provide wireless phones to underserved rural areas in countries, such as India, China, and Russia, where consumers typically have to wait 5 years or more to get local phone service.

The huge potential market for global PCS is the reason some of the world's largest and most sophisticated technology companies are moving ahead with satellite-based phone systems. Most telecommunications analyses see pocket-size satellite phones as the next great mobile communications frontier, with growth that could rival that of the cellular industry. In fact, some analyses predict that the global PCS market will reach $7 billion in annual revenues by 2005.

Phones will cost from $2,000 to $4,000, with charges ranging from about $1 to $3 per minute.

Despite the technical and economic considerations, the global PCS network will soon be a major component of the Information Superhighway.

Bibliography

Brodsky, L., *Wireless: The Revolution in Personal Telecommunications,* Norwood, MA: Artech House, 1995.

Millar, Barry, "Satellites Free the Mobile Phone," *IEEE Spectrum,* March 1998, pp. 26–35.

10 The Future

Television won't hold on to any market it captures after the first six months. People will soon tire of staring at a box every night.

— Darryl Zanuck, head of 20th Century Fox, 1946.

Video-on-Demand (VOD)

In the year 2000, you will switch on your television set, turn to the video channel, key in your preferences and select a current movie using the remote controller.

True VOD services will give you total control. The service provider doesn't begin transmitting the movie until you request it. Once it starts, you can pause, rewind, and fast forward, just like with your VCR.

VOD, home shopping, and interactive games are just some of the interactive services that are expected to be the big money-earners of the Information Superhighway in the near future.

One of the keys to VOD is digital technology. By converting analog television signals into a digital code and using sophisticated digital compression technology, it is possible to reduce the amount of bandwidth or network capacity required to transmit each channel. This means cable companies, and eventually telephone companies, can transmit more channels over their networks.

At the same time, cable and phone companies are rebuilding their networks with fiber optic cable that will increase the total amount of bandwidth. With reduced bandwidth required for each channel and increased total bandwidth, cable and phone companies will be able to deliver hundreds of channels to the home within a couple of years.

MPEG-2, an international standard worked out by the Motion Picture Experts Group, is one key enabling technology. It is a highly sophisticated scheme for compressing video signals. Even so, it will take around 2 GB to store one 2-hour feature film under MPEG-2. That's why digital video servers have thousands of giga-bytes of storage capacity.

A box that sits on top of the television will relay commands from a handheld remote controller to the server and descramble the signal coming back. It will store a few minutes of the movie in computer memory if the pause button is pushed. When you hit play, the movie starts playing from the memory in the set-top box where you left off.

Cable and telephone companies in the United States and Canada are experimenting with VOD and so-called near-video-on-demand systems.

The telecommunications act signed by President Clinton in 1996 now allows a phone company to give the FCC notice of its desire to install a video dial-tone network and so become a common carrier for providers of video programming.

Most of the technical issues of VOD have been resolved. Although debate remains as to the best and most cost-effective technology for delivering interactive television to the home.

Interactive TV Trials

Cablevision operators and telephone companies are testing a number of interactive systems. Some operators are using existing coaxial cable networks plus existing telephone lines for customer input. Some telephone companies are planning to exploit their existing copper pair infrastructure by utilizing the latest digital signal processing techniques. One broadband modem technology is ADSL, which sends compressed digital video over existing twisted copper pairs at up to 9 Mbps. (See Chapter 7.)

Other telcos use a hybrid fiber/coax (H-F/C) network, in which optical fiber lines run out to a remote node (FTTN), and coaxial cable links each home to the node. Others push fiber optics even closer to the home in a configuration known as fiber-to-the-curb (FTTC). This type of network links each home to the fiber by twisted pair copper or coaxial cable.

The various trials differ in the extent of their services. Some offer only limited interactivity, while others like the Time Warner network in Orlando, Florida, are experimenting with a full set of interactive applications.

Time Warner Network

The Time Warner network has a hybrid fiber/coax architecture. Fiber connects central servers to nodes, which serve up to 500 homes. Coaxial cable links each node to a home. The digital set-top box decodes the video streams for display and sends commands from the user back to the server.

At the headend (Figure 10-1) the movie request, for example, is forwarded to a server that stores compressed digital video on disks. The server assembles the data plus the requester's

Figure 10-1 Time Warner Network.

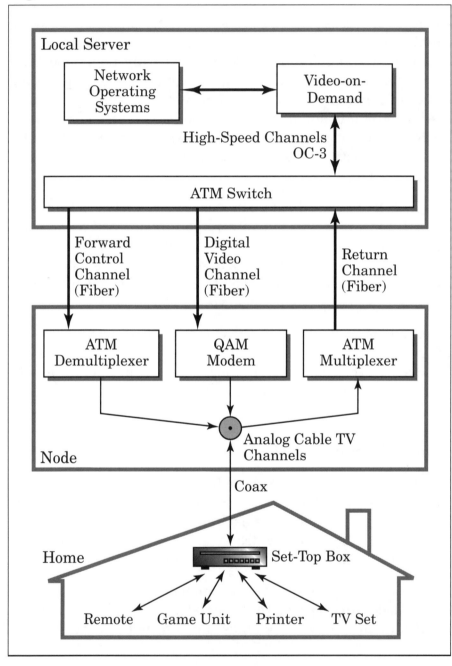

digital address into ATM packets, which are sent to the ATM switch.

On reading the address, the switch routes the packets to the proper frequency modulator and laser for transmission over optical fiber lines to a 500-home node. There the signals are transferred to coaxial cable for the short distance to the customer's home.

The set-top box reassembles the ATM data packets as a data stream, decompresses the video, and displays the movie, which the user may view, fast forward, pause, or rewind.

Although the telephone companies' long-term objective is to provide FTTN or FTTC technology, in the short term they are testing another video delivery system, wireless cable.

Wireless Cable

Wireless cable is a popular oxymoron for multichannel, multipoint distribution system (MMDS). MMDS operates in the 2.1- and 2.7-GHz microwave bands, with a total of 33 analog 6-MHz channels. It has been around for some time, with the frequencies allocated by the FCC some two decades ago for educational use. Most of the licenses are still held by schools and archdioceses, which can lease excess capacity to commercial companies.

Telephone companies became interested in MMDS only in the past few years. That's when compression techniques for digital video improved to the point where digital television became practical.

With digital compression, the 33-analog channel band can transport 100 to 150 digital programs. The rental of the small satellite dish is included in the subscription price.

As shown in Figure 10-2, the signal path for the digital wireless MMDS starts at the super headend, where antennas capture video signals in the 4-GHz frequency range from satellites. At the headend, signal converters digitize, integrate, and compress the signals, which are carried by high-speed data lines to a microwave tower. There it is broadcast as a microwave signal in the 2.7-GHz range to a small antenna on each customer's roof. From the antenna, a coaxial cable carries the digital stream to a set-top box, which converts the digital signal to an analog signal for display on the customer's television set.

The first MMDS digital commercial operation in North America began in 1996 in Brandon, Manitoba, Canada. The system,

Figure 10-2 MMDS.

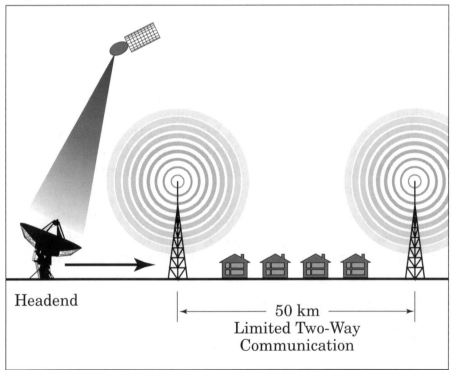

Headend

|←——— 50 km ———→|
Limited Two-Way
Communication

called Sky Cable, uses just nine towers to cover most of the province's populated areas.

A technology similar to MMDS, but with higher potential for interactivity because it is based on a broader bandwidth, is local multipoint distribution system (LMDS). This system operates at around 28 GHz and is similiar to a cellular network. Cell sizes are small to compensate for the shorter transmission distance at that frequency (Figure 10-3). The first full-scale LMDS experiment was started in 1992 in Brighton Beach, New York, by CellularVision.

In August 1998, Nortel Networks shipped the first LMDS equipment to Teligent Inc. in Los Angeles. This was the first commercial LMDS network in North America.

Figure 10-3 LMDS.

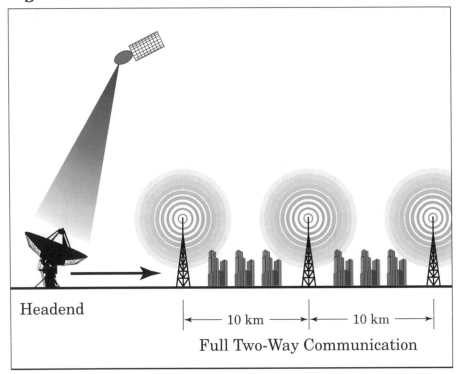

Data Communications

Meanwhile, the cable operators are focusing their attention on data communications rather than interactive television. High-speed access for home computers using cable modems would open up a new world for interactive programming. Although VOD is not yet available, it can provide most of the other likely applications of interactive television, such as home shopping, news services, and games. And unlike interactive television content, much of which has to be specially created, interactive data are already available on the World Wide Web. Rogers Cablesystems, the largest cable operator in Canada, has formed a consortium with other cable firms to provide that service. This new service is called WAVE, and subscribers willing to pay a monthly fee will have high-speed access to the Internet.

The demand for cable modems is growing. Standard telephone modems today have a rate of 28.8 Kbps. With access to ISDN, that rate increases to 64 or 128 Kbps, which is still time consuming for people downloading large files, particularly those with graphics. But attach a modem to a cable line, and communications speed leaps astronomically to as much as 27 Mbps. A multitude of computer users want that increase in modem speed, and will pay for it.

Most cable modems use TDMA within a specified frequency. The cable operator allocates a frequency to a modem channel and can create several such channels. Within each channel as many as 50 modems may operate. It's possible that cable modems, rather than expensive set-top boxes, will bring interactive television to the home.

Digital Television

After more than 10 years of trials, debating, and technical wrangling, the FCC finally approved technical standards for digital television in December 1996. This has come about after an agreement among television manufactures, broadcasters, and representatives of the computer industry, ending a deadlock that threatened to further delay the introduction of high-definition TV (HDTV) in the United States.

With digital television standards in the final stages of approval, computer industry leaders raised objections early in 1996, insisting on modifications to suit computer-driven applications. These could range from incorporating television reception in a personal computer to building computer features into a new generation of hybrid PC-TVs.

The FCC had been reluctant to approve a standard that did not meet the needs of the computer industry, as well as the broadcasting industry. The Clinton administration weighed in on the side of PC manufactures, adding pressure for a compromise. This means U.S. TV viewers can look forward to larger, crisper pictures and CD-quality sound on wide-screen digital TV sets. Standard (analog) TV sets (Figure 10-4) will gradually be displaced by digital HDTV, but U.S. broadcasters will continue analog transmissions for the foreseeable future.

Figure 10-4

In the United States there will be a phase-in period for digital TV beginning in late 1998. Broadcasters will start transmitting their signals on two channels, one in conventional analog and the other in digital. The FCC has assigned, at no charge, second channels for digital broadcasting in parallel with existing analog broadcasts. After a transition period, the broadcasters could return one channel to the government and cease analog broadcasts. By 2006, all TV stations in the United States must be broadcasting in digital and have abandoned analog transmission, or else they risk losing their license.

Figure 10-5

The first of the digital HDTV sets is expected to go on sale in the United States in 1998, with prices that range between $1,500 and $3,000 (Figure 10.5). Prices are expected to come down as sales rise.

The establishment of a U.S. standard for digital TV bodes well for TV manufacturers. Yet it is the PC industry that is expected to set the pace in this emerging market. In the compromise agreement, television industry leaders acquiesced to PC industry demands that the critical "video format" portion of the digital HDTV standard be left open. The market will establish a de facto standard. Personal computers will be challenging TVs in the home entertainment market because PCs will also be able to display digital HDTV (Figure 10-6). This digital convergence means that computer and TV manufacturers will compete and collaborate to develop "hybrid" products that combine computer and digital television technologies (Figure

Figure 10-6

10-7). These hybrid products, as well as combinations of digital televisions with attached set-top boxes (Figure 10-8), are expected to bring interactive technologies such as Internet access, e-mail, computer games, and electronic commerce into U.S. living rooms soon.

Figure 10-7

The dawn of the age of digital television holds the promise of a multibillion-dollar bonanza for TV set manufactures, computer chip companies, glass makers, and digital equipment specialists, and could lead to a merging of the television and personal computer industries.

Figure 10-8

The economy as a whole is sure to benefit as owners of 280 million TV sets throughout the United States replace their sets over the next 10 to 15 years with the new HDTV sets.

Digital Versatile Disc

The introduction of the digital versatile disc (DVD) has revolutionized home theater and multimedia computing.

DVD provides full-length feature films on a single compact disc. The wide-screen pictures have better video quality than LaserDisc. DVD features five discrete channels of digital surround sound, plus a separate subwoofer channel. Sound-tracks will be available in multiple languages as will a viewer rating control, so you can control what your children watch. All major consumer electronics manufacturers now have DVD players available.

The DVD technology provides 133 minutes of better-than-LaserDisc-quality video and digital surround sound on a single CD-size disc. It also provides support for multiple viewing modes, such as wide-screen, letter-boxing, and pan-and-scan (pan-and-scan crops movies made for wide cinema screens so they can be viewed on narrower TV screens).

DVD can achieve horizontal resolution of 720 lines, compared with 400 for LaserDisc, 330 for broadcast TV, and 240 for VHS videocassette. Theoretically, that means more highly detailed pictures. However, with a format like DVD, which uses digital video compression to squeeze as much information as possible onto a disc, horizontal resolution becomes a less meaningful specification.

Because DVD is a digital format, there is virtually no picture noise. That is particularly noticeable in pictures where there is a large expanse of a single color. Current video sources use one signal to record picture brightness and another to record color information. DVD stores component-color signals, which consist of brightness plus separate color component signals for red, green, and blue. The result is more vivid, more accurate colors.

The DVD players sold in North America incorporate Dolby AC-3 digital surround sound, which provides for five completely discrete full-bandwidth channels (front left, front right, and center; left and right surround) plus a subwoofer channel when playing software encoded for AC-3 surround sound. The current standard for surround sound in the home, Dolby Pro Logic, stores audio on two channels, from which it derives center- and surround-channel information.

A standard DVD can hold 4.7 GB, seven times more data than a CD, which holds 650 MB. On its own, the 4.7 GB capacity would not be sufficient to hold 133 minutes of digital video and audio. To hold that much data would require about 160 GB capacity. The DVD uses the MPEG-2 video compression scheme. Video compression achieves the reduction in data by storing and transmitting only the differences between frames. The DVD format supports dual-layer recording, which allows information to be recorded on a semitransmissive layer. More data can be put on a DVD because it has smaller pits that are closer together than a conventional CD. A laser reads the deeper reflective layer, then refocuses and reads the semitransmissive layer (Figure 10-9).

Because DVD players have to be able to read normal CDs as well, a laser mechanism that can change focus is required. Dual-layer discs have a capacity of 8.5 GB. The increased capacity can be used to increase playing time, provide higher bit rates for complex scenes, or have additional soundtracks.

Major film studios have endorsed the DVD format and have released new titles simultaneously on DVD and VHS. While the studios would like DVD to be a sale-only video format, a rental market has developed.

Digital Video Express (Divx)

A new movie format called digital video express (Divx) is now available. A Divx movie is like a conventional DVD, except that it expires after a certain time period and, unlike a rental video, doesn't have to be returned to the store. Instead of picking up a video cassette or DVD at a local video rental store,

Figure 10-9 Digital Versatile Disc (DVD).

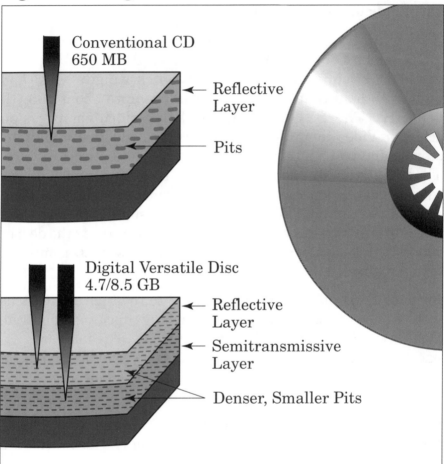

Divx can be purchased in consumer electronic stores for $4.50. Once the disc is slipped into the player, you have a maximum of 48 hours to watch it (as many times as you want). Then the disc locks. At this point, you just throw it away. If you wish to add an additional viewing period, it will cost you $3.25 for another 48 hours. If you wish to have perpetual viewing rights, it can be bought for about $20.00.

Existing DVD players can not play Divx discs, but the new Divx players can play both DVD's and Divx's discs. The Divx players plug into the TV just like DVD players, but they also plug into the phone line. When you first set up your player, you must establish an account with the electronic store. Your machine calls a toll-free number once or twice a month to report on the playing time and bills your credit card.

Computers with DVD-ROMs

Most computers now come equipped with a CD-ROM drive. There are hundreds of CD-ROMs on the market with excellent still graphics, good animation, and CD-quality sound. But video movies on the same CD-ROMs are usually jerky and occupy a small part of the screen. However, full-screen full-motion video is now available for your computer.

Many manufacturers have introduced PCs with MPEG video. MPEG allows audio and video to be stored efficiently on digital media like CD-ROMs. First-generation MPEG technology (MPEG-1) offers picture detail comparable to VHS videocassette. On CD-ROMs with MPEG video, video clips run full-screen instead of inside a small window. On some machines, they run at the same frame rate as broadcast TV.

The quality of full-motion video depends on two factors: resolution (the amount of detail in the picture) and frame rate (the number of images displayed in a second). To deliver a high-resolution image at an acceptable frame rate requires an enormous amount of information to be processed every second.

Showing full-screen full-motion video on a computer monitor means an overwhelming amount of information must be processed, approximately 26 MB of raw data per second! As the miniature size and quality of current video on a computer demonstrates, computers cannot cope with this enormous and con-tinuous data stream. MPEG compresses video and audio sig-nals so they can be efficiently stored on media like CDs and transferred to the computer's processor at reasonable speed. MPEG-1 allows for a data stream of 140 to 180 Kbps. Single-speed CD-ROM has a transfer rate of 150 Kbps with an image size of 352 X 240 at 30 frames per second. MPEG allows for about 75 minutes of video playback from a CD-ROM disc. At this point, the raw MPEG image would still only cover a quarter of the screen. To create the 640 X 480 image, the computer's hardware and software must create additional information. Decoders, either in hardware or software, esti-mate which pixels would be appropriate, based both on the surrounding information in the current frame, as well as information from the previous and following frames.

As PC processors get faster and video cards get smarter, you will be able to play MPEG files without dedicated hardware. Most of the new systems perform MPEG decoding without dedicated hardware. Instead they rely on MPEG decoding software, which exploits the fast Pentium processors and smart graphics cards. Compared to the jerky, partial-screen video movies featured on older multimedia CD-ROMs, MPEG-1 CD-ROMs provide very impressive video quality when played on this new generation of MPEG computers.

With the introduction of DVD, however, there is a second application for this technology. DVDs will be a storage me-dium for personal computers. While CD-ROM made multime-dia computing possible, manufacturers are now constrained by its 650-MB capacity, which at one time seemed infinite. DVD-ROMs can have a capacity of 4.7 GB (for single-layer single-sided discs), 8.5 GB (for dual-layer single-sided discs), or 19.5 GB (for dual-layer double-sided discs). This capacity can be used for multimedia software with more video and more sound.

The next generation of computers will feature complete systems with a built-in DVD-ROM drive and MPEG-2 video playback capability. DVD-ROM multimedia upgrade kits will also be available.

Further down the road are recordable DVDs (DVD-WORM for "write-once read-many") and erasable DVDs (DVD-RAM). Both will, of course, require a new type of drive as well as special media.

Internet Television

This new product will enable users to surf the Web from their living room couch. The sets will have built-in or set-top Internet access.

One product, called WebTV, has developed a platform designed to minimize flicker. This occurs because the electron beam in today's television sets activates alternate rows of phosphor pixels, then returns and fills in the rest at a rate of 60 per second in the United States or 50 per second in Europe. This interlaced scanning works well with the natural scenes of TV broadcasts. But in finely detailed computer images, the alteration of scan lines may be perceptible. In a computer monitor, the beam does not skip lines, but scans progressively. The company's technology is built around a 64-bit 4640 processor from Integrated Device Technology Inc., and operates with Fido, an application-specific IC that acts as a systems controller. Infrared input is received from either a remote control or a wireless keyboard.

WebTV users will gain access to the Internet through the company's own Internet service provider for around $20 a month. Because WebTV created both the service and the product, which requires little software in each unit, the WebTV set-top box has no hard disk and only 2 MB of RAM; most of the storage and application software resides in the network server. WebTV is licensing its technology rather than manufacturing products itself. Philips Electronics and Sony Electronics are the first licensees and began shipping set-top models late in 1996 at a list price under $350.

Competition in this category is expected to heat up quickly. Zenith Electronics has announced its NetVision television, and Thomson Electronics has demonstrated the Genius Theater, a 91-cm color TV with built-in Internet access.

Bibliography

Blackwell, Gerry, "Video-On-Demand," *Home Computing & Entertainment,* January/February 1996, pp. 22–24.

Blackwell, Gerry, "Digital Video Disc," *Home Computing & Entertainment,* March/April 1996, p. 24.

Perry, Tekla S., *The Trials and Travails of Interactive TV,* New York, NY: IEEE Press, April 1996.

11 Summary and Conclusions

Building the Superhighway

The Internet has launched us into a true information age and is enabling people around the world to interact with each other on a scale previously unimaginable. Whatever evolutionary course the Internet and the Information Superhighway takes, the digital convergence of the telecommunications, television, and computer industries has forever altered the way we communicate and compute.

This convergence will enable companies to deliver services that were once beyond their technical limits. Companies envision a world in which they can offer customers one-stop shopping for a bundle of communication, entertainment, and information services that may generate a combined monthly bill well in excess of $100 per household. WebTV will allow viewers to access the Internet. Big screen, high-definition digital television and DVDs may change the way movies are shot. Satellites and on-board computers will help automobiles find their own way home. Households will receive signals from satellites, microwave towers, fiber or copper lines.

A summary of these technologies follows.

- ## **Global PCS Network**

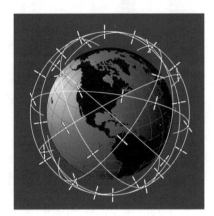

Signals will be bounced from senders to receivers using a network of satellites to be launched around the world by the year 1999.

Four major systems will be available:

Iridium: 66 LEO satellites to transmit voice and data services worldwide.

Globalstar: 48 LEO satellites to supply wireless and other telecommunications service for remote regions and navigation purposes.

Ellipso: 17 LEO satellites to transmit voice and data service worldwide.

ICO Global: 10 MEO satellites to transmit voice and data services worldwide.

Strengths: Unlike wireless and wireline phone calls, signals reach every part of the globe; reliable for standard phone service in less developed countries.

Weaknesses: Expensive; not useful in urban areas and for large corporations with multiple locations.

- ## **Wireline Telephone**

With POTS, with which people are most familiar, calls go along twisted pairs of copper wires from a local phone switch to home or office.

Strengths: A known technology; best voice and sound quality; slower transmission speed improved by ISDN and ADSL technologies. ADSL will be able to deliver broadband transmission capabilities that will allow video and data transmission over conventional phone lines.

Weaknesses: Physical connection can be easily damaged; expensive in less densely populated area; costly base of already installed equipment that can become outdated.

• Wireless Telephone

Types of wireless technology:

Cellular: Older portable technology converts voice signals into radio waves; upgraded digital version turns same voice signals into a series of "on" and "off" pulses.

Personal communication service (PCS): Wireless technology similar to digital cellular service but at a higher frequency and capable of offering many additional features.

Strengths: Portable, not fixed to a particular location; cheaper than wire line in rural areas; advanced PCS handsets will have capability of accessing the global satellite network.

Weaknesses: Capacity limitations addressed by a series of engineering techniques such as CDMA; air signals face interference; calls can be abruptly ended without warning; expensive to send data signals.

- ## Cable

  H-F/C cable networks already partially installed delivers television signals and telephone service.

  Strengths: Wired link provides reliability; provides the high bandwidth necessary to deliver video, telephone, and future interactive services including VOD.

  Weakness: High cost to upgrade cable TV networks to provide two-way capability for telephone service.

- ## Direct-to-Home (DTH) Satellite Television

  From a single satellite in the sky, DTH can deliver hundreds of digital television channels to pizza sized dishes in urban, rural, and remote areas.

  Because costs are only a fraction of a large dish satellite television, this product has become the fastest growing consumer electronic product ever introduced in the United States.

  Strengths: Ability to provide many channels of television signals from a single source.

  Weaknesses: No local channels in some cases; service is entirely dependent on satellites that are expensive to build and launch, have a lifespan limited by their fuel capacity, and have occasional, unfixable failures.

• Multichannel Multipoint Distribution System (MMDS)

Also known as wireless cable or microwave television, this is the latest generation of traditional lower frequency broadcast technology. It operates in the 2.1- and 2.7-GHz bands. Tall communication towers broadcast signals in a radius of about 25 km.

Strengths: Low-cost distribution of television signals; costs are largely related to customers receiving equipment rather than network infrastructure, so costs increase only as the number of subscribers grows; able to carry local signals.

Weaknesses: Limited two-way capability because other broadcast activities in the lower frequency ranges around the 2.5-GHz band limit available bandwidth for MMDS to offer a full range of services.

• Local Multipoint Distribution Systems (LMDS)

Technological advances mean this new broadcast technology can operate in ultra-high frequency bands that previously were considered unusable.

LMDS frequencies are in the 28-

GHz range. Signals broadcast from highrise apartment tow-ers and office buildings have a radius of just 5 km. Because 1,000 MHz of frequency has been allocated to LMDS, it will be able to deliver a full range of video, voice, and data sig-nals including such things as videoconferencing and high-speed Internet access as well as traditional cable television and phone service. Sometimes called "virtual fiber."

Strengths: Wide bandwidth will allow operators to pro-vide a full range of services such as video downloading by using television remote controls.

Weaknesses: Requires a much greater number of expen-sive transmission sites than MMDS; technology is so new that there are few commercial systems operating in the world.

The Players

The major players in this digital convergence game are the telephone companies and the cablevision companies. Both players are spending enormous sums of money to create the infrastructure for the Information Superhighway. Some of the technologies they intend to use are still in their infancy, and because the cost of the infrastructure is so expensive, both play-ers are constantly reviewing their strategy. As a result, each sector is considering different solutions to connect the consumer to the Superhighway.

Both sectors would like to offer the consumer a total package of services that would include improved telephone service, television, interactive television (e.g., video-on-demand, on-line banking), and high-speed Internet access. The cost to the con-sumer will depend on the choice of technologies, especially for the "last mile" of wiring that connects the consumer to the network. The stakes are very high and the total cost is a moving target, so there will be both winners and losers.

Telephone Company Strategies

The telephone companies can now compete in the television market. They could add more H-F/C cable to replace the copper pairs, but that could cost up to $3,000 per customer. A better alternative might be to provide ADSL which is designed to operate over existing copper pairs and provide additional bandwidth. Another alternative would be to provide MMDS or LMDS. Many telephone companies are planning to use these technologies as a means of entering the market sooner with less capital cost. A longer term strategy could be a direct broadcast satellite service. Based on the experience of the direct-to-home satellite operators, the telephone companies will need at least two million subscribers to break even.

The existing Internet access service provided by the telephone companies will be upgraded over time, but the appropriate technology for the long term has yet to be decided. As an option to the standard telephone modem speeds of 14.4 and 28.8 Kbps, ISDN lines are currently available at 64 to 128 Kbps. Many telephone companies, however, are planning to use ADSL at 1.5 Mbps to provide higher access speeds to the Internet. Their long-term objective would be to provide FTTH or FTTC technology.

Cable Company Strategies

The cable companies will be supplying cable modems for high-speed access to the Internet at speeds up to 27 Mbps, for customers willing to pay higher rates. However, if the computer server at a Web site runs at a slower speed than your modem, your link will default to the slower speed. Some cable companies plan to offer a catching service that will store frequently accessed Internet sites on their own high-speed servers, allowing their subscribers to access the site at high speed. All of these upgrades will be costly because the cable companies have to convert their existing one-way coaxial/fiber cables networks to two-way interactive access capable of transmitting data and video. This requires the installation of bidirectional amplifiers and other equipment upgrades. It remains to be seen whether consumers will be willing to pay for VOD and high-speed Internet access in order to justify the companies' cost. Some companies may choose to provide near-video- on-demand, which is really an extended form of pay-per-view, as well as some Internet access via their television set, which could include on-line banking and shopping. In addition to installation fees, the subscriber would have to buy or rent a set-top box that performs the digital compression of the video signals. There may not be sufficient incentive for some companies to make the heavy investment in interactive TV.

Some cable companies are also anxious to get into the multibillion dollar local telephone market. This will also require extensive changes to their cable facilities. The telephone in the customer's house will have to be connected through an interface unit to the coaxial cable and a backup power supply to support the telephone service if the power goes out. The service will also require the bidirectional amplifiers that are necessary for interactive TV.

The cable companies hope to capture at least 15% of the local phone market. But the big question is whether the customer will have enough faith in the local cable companies to trust them with something as essential as local phone service. The cable companies' main problems go beyond the technological.

They must find new sources of revenue because their market is not growing very fast and is threatened by the new competition. The main threat is the telephone companies' ability to bundle together all the new services and deliver them as a package to the customer. Also, just because something is technologically possible doesn't mean its economically viable.

Early in 1997, Tele-Communications Inc. (TCI), one of the largest cable companies in the United States, admitted that their vision for TCI's future as a multimedia powerhouse, straddling television, telephones, and the Internet, wasn't working. It was too ambitious, overhyped, and impossible to carry out on schedule. In its place, TCI is pursuing a much diminished strategy and is making a major retreat to its roots in cable.

The telephone companies, on the other hand, can raise capital for new businesses much more easily. But they must become full-service providers. It's the only way they can reduce churn (customers changing providers). In addition, the major telephone companies are facing competition in their lucrative long distance markets from new international operators offering wireless PCS via satellite.

The Internet

In the telecommunications industry, digital and fiber optic technologies have over the past 15 years doubled in transport capacity every 16 to 24 months. The result has been an exponential reduction in the cost of network bandwidth. The telecommunications and computer industries have each progressed down separate, stable, and predictable evolutionary paths. Today, the paths of both industries have converged dramatically, fusing their future directions and, in the process, taking their planned evolution significantly off course. At the same time, their convergence is unleashing an innovative force that is changing the way people communicate at the most fundamental levels.

The recent advent of two simple but clever pieces of software, the World Wide Web and network browsers, quickly transformed the Internet into a multimedia structure, sweeping away the use of archaic commands and the need to possess detailed technical knowledge. Suddenly, the Internet became accessible to everyone, not just to surfers and technologists. As a result, its value soared. The latest generation of networking terminology promises to propel the Internet to a still wider audience. Network programming languages such as Java are causing an exciting new paradigm for cost-effective, platform-independent software development and delivery via the World Wide Web. And the network computer, essentially a computer without capabilities until it is connected to a network, will deliver a scaled down, lower cost appliance optimized for distributed computing.

In the Internet market, the cable companies have the lead in speed, and it will be a few years before the phone companies can compete on this level for the whole network. However, the average home will always choose price over speed. Only the true surfer will pay a premium price for high-speed access. A further concern to both cable and telephone companies is that there's more hype than profit to the Internet. Many business customers are unhappy with Internet security, convenience, and capability. Until the real wideband, interactive Information Superhighway is completed, the growth of the Internet will be limited.

Wireless Technology

Wireless technology is growing at a remarkable rate, and companies are betting billions of dollars on this new technology. The new wireless phone networks of digital cellular and PCS use three different standards: CDMA, TDMA, and GSM. Backing the wrong format could be expensive. As with the VHS-BETAMAX video debate of the 1970s, the companies must pick a format and run with it.

In the United States, CDMA has captured 51% of the market, followed by TDMA-based systems at 40%. If CDMA catches on, those companies that guessed correctly will be able to operate their new pocket telephone systems more cheaply and offer more new calling services. It they're wrong, they'll have spent a lot of money for only marginal improvement in the network and may have set themselves up for some serious problems.

But until digital technology overtakes analog technology and while so many different standards of PCS exist, there will always be the problem of compatibility. Manufacturers of PCS handsets already have some solutions planned. One approach is a dual-mode phone. If you are talking on a digital network and roam to an area without digital, it would switch automatically to analog. There will also be a system for using PCS phones in countries with incompatible systems. A small chip in the back of the phone could be removed and replaced with one of the appropriate operating system. There is also talk of a world phone, or multifrequency phone, that would operate in any area.

Home Entertainment

About 110 million television receivers will be sold in 1998, as well as some 90 million personal computers. As the two begin their battle to become the hardware basis for future home entertainment systems, the question is whether one or the other may eventually come to dominate, or whether the two may yet call a truce and merge into a single unit. As for the television set itself, flat panel displays are coming. Several manufacturers have introduced 106-cm diagonal gas plasma displays, 10 cm thick.

Fujitsu recently announced that it had developed a technology called *alternate lighting of surfaces*, for high-resolution plasma display panels. The technology allows displays of upward of 31 million dots, and more than 1,000 scan lines, more than double the 480 scan lines used in current plasma display panels.

The new DVD is very important both for video (because a single disc can hold a full-length movie) and for storing information for display on a personal computer. Some speculate that the annual market for DVD players could reach 80 million units by 2000.

Fiber Optic Technology

Using light to transmit information is not a new idea. More than a hundred years ago, Alexander Graham Bell transmitted a telephone signal over a distance of 200m using a beam of sunlight as the carrier. That historic event involving the "photophone" marked the first demonstration of the basic principles of optical communications as it is practiced today. The photophone did not reach commercial fruition, however, due to the lack of a reliable, intense light source and a dependable, low-loss transmission medium.

Today, optical fiber is the backbone of long distance telecommunications and is now moving into the local loop to serve homes and businesses. In the 1970s, when the telephone companies in the United States and Canada began to field trial experimental fiber optic transmission systems, transmission was restricted to one channel or wavelength for each fiber.

Today, using WDM techniques, we can use up to 80 wavelengths on a single fiber with each wavelength transmitting 2.5 Gbps. This means that a 1-inch cable containing up to 432 fibers can transmit at 173 Tbps.

The key to breaking the terabit barrier has been the ability to remove impurities, especially water, from fiber. As a result, the transmission window is now between 1,300 and 1,700 nm in the infrared band. Eventually, it may be possible to provide 5,000 wavelengths over one fiber.

The final frontier is optical switching. Research is continuing using optical mirrors that will add or drop specific wavelengths without disturbing the remaining channels. While fiber optic technology is providing limitless bandwidth, computer processing speed is not keeping up. Projections indicate that there will be 60 million dial-up connections for Internet access in North America by 2002.

This potential bottleneck will disappear only when fiber optic systems are installed in the so-called last mile of the network, the local loop between the central office and the customer's home or business.

Construction Underway

Despite the technical problems facing the industry, the global Information Superhighway will be built. Large portions have already been completed. The telecommunications industry is currently very profitable. The ITU estimates revenues for the global business, including the computing and audio-visual sectors, at \$1.43 trillion, equivalent to 5.9% of the world's gross domestic product. The gross operating margin, a measure of profitability, is estimated to be 40% of revenues. However, the growth of the basic business of providing telephone lines and services is falling rapidly, at least in developed countries. New competition is eating away at profit margins; for many companies, it is easier to buy a competitor with a promising project or service than to develop their own. This is the principal cause of the rash of mergers and strategic alliances now convulsing the global industry. Telecom operators must find new products, services, and markets to exploit.

The completion of the global Information Superhighway will lead to the development of products, services, and industries that are beyond our imagination today.

On the Information Superhighway, construction has begun.

Telecommunications Glossary

Add/drop multiplexer (ADM)
A multiplexer that will allow individual channels to be dropped at nodes of a network and other channels to be inserted in their place.

Advanced Mobile Phone System (AMPS)
An analog cellular radio system used widely in the United States.

Africa ONE
A 40,000-km fiber optic network that encircles the entire continent of Africa.

American National Standards Institute (ANSI)
Approves and promulgates standards developed by accredited U.S. standards committees. Also coordinates standards with international standards organizations and manages U.S. participation in international standards activities.

Amplifier
A device for increasing the power of a signal.

Amplitude
The amount of variation of an alternating waveform from its zero value.

Amplitude modulation (AM)
The process by which a continuous waveform is caused to vary in amplitude by the action of another wave containing information.

Analog

In communications, the description of the continuous wave or signal (such as the human voice) for which conventional telephone lines are designed.

Asymmetric digital subscriber line (ADSL)

A transmission system that will deliver a digital channel at a rate up to 9 Mbps in one direction over copper pairs, depending on distance. It will deliver lower speed channels at distances up to 18,000 feet. It will also provide a POTS channel in both directions, plus a low speed (16-Kbps) control channel.

Asynchronous transfer mode (ATM)

A system that transports data over a network in uniform 53-byte packets or cells. The packets contain a 5-byte address and 48 data bytes. Also known as "cell relay."

Asynchronous transmission

A transmission method in which each character of information is individually synchronized, usually by the use of "start" and "stop" elements (compare with synchronous transmission).

Attenuation

A decrease in the power of a signal while being transmitted between points. Usually measured in decibels.

Automatic call distribution (ACD)

A system that will distribute incoming telephone calls and direct them to the department that can provide the service desired by the caller. The callers are presented with voice menus of options, which they can select by pressing buttons on a touch-tone telephone.

Bandwidth

The difference between the high and low frequencies of a transmission band, expressed in Hertz.

Baseband

In analog terms, the original bandwidth of a signal from a device (i.e., 4 kHz for telephone, 4.5 MHz for television).

Baud

A unit of signaling speed. Speed as expressed in bauds is equal to the number of signaling elements per second. At low speeds (under 300 bps), bits per second and baud are the same. But as speed increases, baud is different from bits per second because several bits are typically encoded per signal element.

Binary

Having only two possible states.

Bit

A contraction of binary digit; the smallest unit of information in a binary system of notation. Data bits are used in combination to form characters; framing bits are used for parity, transmission synchronization, and so on.

Broadband (wideband) channel

A communication channel with a bandwidth larger than that required for baseband transmission. Very often any channel wider than voice grade is considered to be a broadband channel.

bps

Bits per second (also expressed as bps) is a measure of speed in serial transmission. Also used to describe hardware capabilities, as in a 9,600 b/s modem.

Byte

A group of 8 bits. Often used to represent a character. Also called an octet.

Cable TV

Successor to CATV. Distribution of multichannel television programming throughout an area by a cable-TV network operator.

C band

Microwave radio frequency band approximately 4 to 6 GHz.

Central office

A place where telephone circuits are switched automatically, to connect a calling station to a called station.

Channel

In communications, a path for transmission (usually one way) between two or more points. Through multiplexing, several channels may share common equipment.

Character

Any coded representation of an alphabet letter, numerical digit, or special symbol.

Clock

In data communications, a device that generates precisely spaced timing pulses (or the pulses themselves) used for synchronizing transmissions and recording elapsed times.

Coaxial cable

A cable that consists of an outer conductor concentric with an inner conductor; the two are separated from each other by insulating material.

Code

A specific way of using symbols and rules to represent information.

Codec

Codec stands for coder/decoder. An electronic device which converts analog signals to digital form, and back again.

Codec, video

A codec that converts an analog video signal to digital form, and back again. In addition, video codecs generally compress or reduce the data rate to that which can be carried on a narrowband channel.

Code division multiple access (CDMA)

A transmission technology that distributes a signal over a broad bandwidth by combining it with a high bit rate pulse stream selected in accordance with a pseudo-random code. A receiver can recover the original signal when provided with the same pseudo-random code.

Community antenna television (CATV)

Reception of broadcast television signals in an area or a community with marginal television reception, using a common antenna (usually located at a good signal site) and distribution of the received channels throughout the area over coaxial cable.

Competitive access provider (CAP)

A telephone company that provides local exchange service in competition with established local telephone providers. This service was established by the Tele-communications Act of 1996.

Computer/telephony integration (CTI)

Integration of telephone service and computer facilities to provide faster response to callers' inquiries.

CPU

Central Processing Unit – the part of a computer that includes the circuitry for interpreting and executing instructions.

Cross-connect

Equipment for connecting channels from incoming trunks to outgoing trunks to direct them to the desired office.

Crosstalk

Interference or an unwanted signal from one transmission circuit, detected on another (usually parallel) circuit.

CRT

Cathode Ray Tube – an electronic vacuum tube, such as a television picture tube, that can be used to display images.

Data

A representation of facts, concepts, or instructions in a formalized manner suitable for communication, interpretation, or processing; any representations, such as characters, to which meaning may be assigned.

Data collection

The act of bringing data from one or more points to a central point.

Data communications

The movement of encoded information by an electrical transmission system. The transmission of data from one point to another.

Dataphone

A trademark of the AT&T Company to identify the data sets manufactured and supplied by it.

Data processing

The execution of a systematic sequence of operations performed upon data.

Data processing system

A network of machine components capable of accepting information, processing it according to a plan, and producing the desired results.

Data set

A device containing the electrical circuitry necessary to connect data processing equipment to a communications channel, usually through modulation and demodulation of the signal.

Data sink

The equipment that accepts the transmitted data.

Data source

The equipment that supplies the data signals to be transmitted.

Data stream

Generally, the flow of information being transmitted in a communications system or path.

dBm

A measure of power in communications: the decibel referenced to 1 milliwatt: 0 dBm = 1 milliwatt, with a logarithmic relationship as the values increase or decrease.

DDS

Dataphone® Digital Service, or dedicated digital service, a private-line service for digital data communications.

Decibel (dB)

A unit for stating the logarithmic ratio between two amounts of power. It can be used to express gain or loss without reference to absolute quantities.

Dedicated link

A leased telephone line, reserved for the exclusive use of one customer; private line; special access.

Demodulation

The process of retrieving an original signal from a modulated carrier wave. The technique used in data sets to make communication signals compatible with business machine signals.

Detector

The means (usually PIN or APD) used to convert an optical signal to an electrical signal.

Diagnostics

The detection and isolation of a malfunction or mistake in a communications device, network, or system.

Digital

In data communications, the description of the binary (off/on) output of a computer or terminal. Modems convert the digital signals into analog waves for transmission over conventional analog telephone lines.

Digital versatile disc (DVD)

Discs on which entertainment material such as movies, concerts, and photographs are optically recorded in digital form. Digital versatile disc drives are available for most computers.

Direct broadcast satellite (DBS)

Satellites that broadcast multiple television channels directly to antennas of subscriber receivers.

Dispersion

Spreading of the light pulses in a digital signal stream of a fiber optic transmission system due to delay variations of different modes or light wavelengths by the optical fiber.

Distortion

The unwanted change in waveform that occurs between two points in a transmission system.

Distributed processing

A general term usually referring to the use of intelligent or programmable terminals for processing at sites remote from a company's main computer facility.

EIA interface

A standardized set of signal characteristics (time duration, voltage, and current) specified by the Electronic Industries Association.

Emitter

The means (usually LED or laser) used to convert an electrical signal into an optical signal for transmission by an optical waveguide.

Erbium doped optical fiber amplifier

An optical amplifier utilizing a section of optical fiber doped with the rare earth erbium and optically pumped with a laser diode. It can amplify a range of wavelengths at the same time.

Error rate

The ratio of incorrectly received data (bits, elements, characters, or blocks) to the total number transmitted.

Facsimile (also called fax)

The transmission of photographs, maps, diagrams, and other graphic data by communications channels. The image is scanned at the transmitting site, transmitted as a series of impulses, and reconstructed at the receiving station to be duplicated on paper.

Federal Communications Commission (FCC)

A board of commissioners that regulate all interstate and foreign electrical communications originating in the United States.

Feedback

The return of part of the output of a machine, process, or system to the input, especially for self-correcting or control purposes.

Fiber

A single, separate optical transmission element, characterized by a core and a cladding.

Fiber-in-the-loop (FITL)

Optical fiber from a telephone central office to a pedestal or to a user premise to provide distribution of communication services.

Fiber optics (lightwave)

Light transmission through optical fibers for communication or signaling.

Fiber optic link around the globe (FLAG)

A 27,000 km fiber optic cable network from England to Japan, with landing points in Europe, the Middle East, Africa, and Asia.

Four-wire circuit

A circuit containing two pairs of conductors, one pair for the transmit channel and the other for the receive channel. A communication path in which there are two wires for each direction of transmission.

Frame

A complete video image, consisting of two fields. Each NTSC frame is made up of 525 scan lines, half of which are allocated to each field. For full-motion video, frames are transmitted at the rate of 30 per second. The European PAL standard dictates frames of 625 scan lines sent at the rate of 25 per second.

Frequency

The number of cycles per unit of time, denoted by Hertz (Hz).

Frequency division multiplex

A system of transmission in which the available frequency transmission range is divided into narrower bands, so that separate messages may be transmitted simultaneously on a single circuit.

Frequency modulation (FM)

A method of modulation in which the frequency of the carrier is varied according to the amplitude of the transmitted signal.

FSK

Frequency shift keying—the most common form of frequency modulation, in which the two possible states (1/0, on/off, yes/no) are transmitted as two separate frequencies.

Full duplex

Used to describe a communications system or component capable of transmitting and receiving data simultaneously.

Geostationary orbit (GEO)

A stable satellite orbit wherein the centripetal force due to the orbital velocity counterbalances the gravitational force of the Earth. At the required orbital velocity and distance from the Earth, the satellite remains above the same point on Earth and appears stationary.

Gigabits per second (Gbps)

One billion bits of digital information transmitted per second.

Gigahertz (GHz)

A unit of frequency equal to one billion Hertz.

Global Positioning System (GPS)

A group of 24 satellites placed into orbit by the United States. The satellites broadcast codes that enable receivers on Earth to calculate their location.

Global System for Mobile Communications (GSM)

A standard for digital cellular telephones that was developed in Europe and has been adopted by many countries around the world.

Half-duplex

Used to describe a communications system or component capable of transmitting data alternately, but not simultaneously, in two directions.

Hertz (Hz)

Synonymous with cycles per second: a unit of frequency, 1 Hz is equal to one cycle per second.

High data rate subscriber line (HDSL)

HDSL provides symmetrical transmission at a data rate of 1.536 Mbps in both directions over a single pair of copper wires.

High-definition television (HDTV)

A broadcast television system that calls for the transmission of 1,125 line frames, developed by NHK, the Japanese broadcasting company.

Hybrid

A bridge-type device used to connect a 4-wire line to a 2-wire line so that both directions of transmission on the 4-wire line are isolated from each other, but are connected to the 2-wire line.

Hybrid fiber/coax (H-F/C)

A distibution system for cable television that employs fiber optic cable to transport multiple TV channels in a 50- to 750-MHz composite radio frequency signal to optical/electrical converters at distribution nodes. The signal is carried from the nodes to individual premises over coaxial cables.

IEEE

The Institute of Electrical and Electronic Engineers—a publishing and standards-setting body responsible for many standards in the communications industry.

Inductance

A measure of the ability of a coil of wire, called an inductor, to block high-frequency signals from flowing through it. Inductance is measured in units of Henrys (H).

Information bit

A bit used as part of a data character within a code group (as opposed to a framing bit).

Infrared

The band of electromagnetic wavelengths between 0.75 micron and 1,000 microns.

Integrated services digital network (ISDN)

Digital network with a basic rate interface that provides a user with two 64-kbps channels and a 16-kbps signaling channel. The primary rate interface provides twenty-three 64-kbps channels and one 16-kbps signaling channel.

Intelligent terminal

A "programmable" terminal that is capable of interacting with the central site computer and performing limited processing functions at the remote site.

Interactive voice response (IVR)

Used for CTI and ACD systems. The callers are presented with voice menus of options, which they can select by pressing buttons on a touch-tone telephone.

Interexchange carrier (IXC or IEC)

A carrier that provides interconnection between two different service areas; usually a toll carrier.

Interface

A shared connection or boundary between two devices or systems. The point at which two devices or systems are linked. Common interface standards include EIA Standard RS-232B/C, adopted by the Electronic Industries Association to ensure uniformity among most manufacturers.

International Telecommunications Union (ITU)

An agency of the United Nations, headquartered in Geneva, Switzerland, to carry out studies of world telecommunications and make recommendations for standardization.

Kilohertz (kHz)

A unit of frequency equal to 1,000 Hz.

Kilometer (km)

1,000 meters or 3,281 feet (0.621 mile).

Ku band

A microwave radio frequency band around 15 GHz.

Laser

An acronym for light amplification by stimulated emission of radiation. A source of light with a narrow beam and a narrow spectral bandwidth.

Leased Channel

A point-to-point channel reserved for the sole use of a single leasing customer. (See dedicated link.)

Light emitting diode (LED)

A semiconductor device that emits incoherent light.

Lightwave
See Fiber optics.

Local multipoint distribution system (LMDS)
An interactive wireless transmission system for distribution of television programs and high-speed access to the Internet. LMDS operates in a 27.5- to 29.25-GHz frequency band.

Loopback tests
A test procedure in which signals are looped from a test center through a modem or loopback switch and back to the test center for measurement.

Low Earth orbit (LEO)
An orbit at an altitude around 1000 km.

Medium Earth orbit (MEO)
An orbit at an altitude around 10,000 km.

Megabits per second (Mbps)
One million bits of digital information transmitted per second.

Megahertz (MHz)
Unit of frequency equal to one million Hertz.

Metropolitan area network (MAN)
A group of local area networks (LANs) connected together over a distance of up to 50 km.

Micron (μm)
Micrometer. Millionth of a meter = 10^{-6} meter.

Microsecond (μs)
One millionth of a second = 10^{-6} second.

Microwave
Any electromagnetic wave in the radio frequency spectrum above 890 MHz.

Milliamperes (mA)

The measure of the current flowing in an electrical circuit. One thousandth of an amp. 1,000 mA = 1 A

Millisecond (ms)

One thousandth of a second.

Mode

A method of operation (as in binary mode).

Modem

A contraction of modulator/demodulator.

Modulation

The process whereby a carrier wave is varied as a function of the instantaneous value of the modulating wave.

Motion Picture Experts Group (MPEG)

A group that develops standards for video transmission, including standards for video compression.

MPEG-2

An algorithm for compression of broadcast quality video.

MSAT©

A satellite that can provide communications directly with mobile, vehicular, or handheld access units. MSAT operates in the 1- to 2-GHz frequency band.

Multichannel multipoint distribution system (MMDS)

An interactive transmission system that can provide distribution of 136 channels of compressed video, including 40 channels of near-video-on-demand, and two-way high-speed Internet access.

Multimode fiber

An optical waveguide that allows more than one mode to propagate. Either step-index or graded-index fibers may be multimode.

Multiplex

> To interleave or simultaneously transmit two or more messages on a single channel.

Multiplexing

> The process of dividing a transmission facility into two or more channels.

Multipoint Circuit

> A circuit that interconnects three or more stations.

Nanometer (nm)

> One billionth of a meter = 10^{-9} meter.

Nanosecond (ns)

> One billionth of a second = 10^{-9} second.

National information infrastructure (NII)

> The Information Superhighway, a telecommunications network that will allow people, agencies, governments, and communities to exchange voice, data, and video information.

Network

> A series of points interconnected by communications channels. The switched telephone network consists of public telephone lines normally used for dialed telephone calls; a private network is a configuration of communications channels reserved for the use of a sole customer.

Network management system

> A comprehensive system of equipment used in monitoring, controlling, and managing a data communications network. Usually consists of testing devices, CRT displays and printers, patch panels, and circuitry for diagnostics and reconfiguration of channels, generally housed together in an operator console unit.

Noise

Generally, any disturbance that interferes with the normal operation of a communications device or system. Random electrical signals, introduced by circuit components or natural disturbances, which degrade the performance of a communications channel.

NTSC

National Television Standards Committee. The North American standard for color television systems. Calls for 525 line frames transmitted at the rate of 30 per second.

Numbering plan area

A geographical area that is assigned a unique area code in accordance with the North America Numbering Plan.

Octet (See Byte.)

Ohm

The unit of measurement for the resistance (DC) and impedance (AC) of an electrical circuit.

On-line system

A system in which the data to be input enters the computer directly from the point of origin (which may be remote from the central site) or the output data are transmitted directly to the location where they are to be used.

Packet

A group of bits, including address, data, and control elements that are transmitted and switched together.

Packet switching

A data transmission method, using packets, whereby a channel is occupied only for the duration of transmission of the packet.

Parity check

A checking system that tests to ensure that the number of 1s or 0s in any array of binary digits is con-sistently odd or even. Parity checking detects characters, blocks, or any other bit grouping that contains single errors.

Passive optical network (PON)

A branching network composed of optical fibers and passive optical splitters. Passive optical networks are used in fiber-in-the loop (FITL) systems to distribute communications from remote terminals to subscribers.

Petabits per second

One quadrillion (10^{15}) bits of digital information transmitted per second.

Petahertz (PHz)

Unit of frequency equal to one quadrillion Hertz.

Phase Alternating Line (PAL)

The broadcast television standard in Europe calling for 625 lines per frame transmitted at 25 frames per second.

Photonics

A technology based on interactions between electrons and photons.

PIN diode

A device used to convert optical signals to electrical signals in a receiver.

Plain old telephone service (POTS)

Dialed voice frequency channel telephone service.

Point of presence (POP)

A meeting point, within a service area, between a local exchange carrier (LEC) and an interexchange carrier (IXC). A POP is a physical location, usually in a building.

Polling

A centrally controlled method of calling a number of terminals to permit them to transmit information. As an alternative to contention, polling ensures that no single terminal is kept waiting for as long a time as it might under a contention network.

Postal, Telegraph, and Telephone (PTT)

Government organization that, until recently, functioned as the common carrier in many countries.

Pulse Code Modulation (PCM)

A modulating analog signal is sampled, quantized, and coded so that each element of the information consists of different kinds or numbers of pulses and spaces.

Real time

Generally, an operating mode under which receiving the data, processing it, and returning the results takes place so quickly as to actually affect the functioning of the environment, guide the physical processes in question, or interact instantaneously with the human user(s). Examples include a process control system in manufacturing or a computer-assisted instruction system in an educational institution.

Refractive index

The ratio of light velocity in a vacuum to its velocity in the transmitting medium.

Regional Bell operating company (RBOC)

Companies created when AT&T divested itself of its operating telephone companies in 1982. There were initially seven, but two pairs have recombined so that there are now five.

Repeater

A device in which signals received over one circuit are

automatically repeated in another circuit or circuits, generally amplified, restored, or reshaped to compensate for distortion or attenuation.

Response time

The time a system takes to react to a given act; the interval between completion of an input message and receipt of an output response. In data communications, response time includes transmission times to the computer, processing time at the computer (including access of file records), and transmission time back to the terminal.

Serial transmission

A mode of transmission in which each bit of a character is sent sequentially on a single circuit or channel, rather than simultaneously, as in parallel transmission.

Signaling System Number 7 (SS7)

An International Telecommunications Union (ITU) standard for out-of-band interexchange signaling.

Simplex

Generally, a communications system or device capable of transmission in one direction only.

Singlemode fiber

A fiber waveguide on which only one mode will propagate, providing the ultimate in bandwidth. It must be used with laser light sources.

Specialized common carrier (SCC)

A common carrier that specializes in a particular telecommunications service, such as long distance.

Spectral bandwidth

The difference between wavelengths at which the radiant intensity of illumination is half its peak intensity.

Switched telephone network

A network of telephone lines normally used for dialed telephone calls. Generally synonymous with the direct distance dialing network, or any switching arrangement that does not require operator intervention.

Synchronous digital hierarchy (SDH)

ITU-T international standard for digital transmission over optical fiber.

Synchronous optical network (SONET)

ANSI standard for digital transmission over optical fiber. Compatible with SDH at 155.52 Mbps and higher rates.

Synchronous transmission

A transmission method in which the synchronizing of characters is controlled by timing signals generated at the sending and receiving stations (as opposed to start/stop communications). Both stations operate continuously at the same frequency and are maintained in a desired phase relationship. Any of several data codes may be used for the transmission, so long as the code utilizes the required line control characters (also called bi-sync or binary synchronous).

Telegraph

An apparatus for communicating at a distance by means of prearranged signals or codes.

Teleprinter/teletypewriter

A machine that can send coded electrical signals in response to a keyboard input and print alphabetical and numeric characters in response to similarly received signals. Teletype is a trademark of the Teletype Corporation.

Terabits per second (Tbps)

One trillion bits per second.

Terahertz

Unit of frequency equal to one trillion Hertz.

Terminal

Any device capable of sending and/or receiving information over a communication channel, including input to and output from the system of which it is a part. Also, any point at which information enters or leaves a communication network.

Time division multiple access (TDMA)

A digital cellular standard. This technique is also used in satellite communications and allows more than one Earth station access to a single satellite channel.

Time division multiplexer

A device that permits the simultaneous transmission of many independent channels into a single high-speed data stream by dividing the signal into successive alternate bits.

Transatlantic Telephone Cable (TAT)

Telecommunications cables between the United States and Europe. Series began with TAT-1. TAT-14 is currently being installed.

Transmission Control Protocol/Internet Protocol (TCP/IP)

A set of data communications standards used initially for the Internet to interconnect dissimilar networks and computing systems.

Transpacific Cable (TPC)

A transpacific self-healing ring network between Japan and the United States with landing sites in Guam and Hawaii.

Trunk

A single circuit between two points, both of which are switching centers and/or individual distribution points.

Voice grade channel

A channel suitable for the transmission of speech, digital or analog data, or facsimile, generally having a frequency range of about 300 to 3,000 Hz.

Wavelength division multiplexing (WDM)

A technique that employs more than one light source and detector operating at different wavelengths and simultaneously transmits optical signals through the same fiber while message integrity of each signal is preserved.

Wide area network (WAN)

A telecommunications network that covers a large geographic area. It typically links cities and may be owned by a private corporation or by a public telecom operator.

Wideband Channel

A channel broader in bandwidth than a voice-grade channel.

Wire telegraph

A system employing the interruption or reversal of polarity of a direct current electrical circuit, in accordance with a code, for communicating at a distance.

X.25

ITU's international standard that defines the interfaces between a packet-mode user device and a public data network.

X.75

ITU's international standard for connecting packet switched networks.

Cyberspace Glossary

Address

Secret code by which the Internet identifies users.

Archie

A system that tracks information anywhere on the Internet.

Bandwidth

Number of bits of information that can move over a communications medium in a given amount of time.

BBS

For bulletin-board system; a congregation gathered (electronically) around a modem that allows users to post messages; they began as informal communities but now include political and even commercial categories.

Beta

In the preliminary, or testing, stage; as in "they're still in the beta mode with that software."

Browser

A program that allows a user to navigate through nodes of the Internet.

CD-ROM

Read only memory—a laser-encoded disc that stores 650 MB of randomly accessible text, imagery, and/or sound data.

Cracker

Hacker who breaks a computer system.

Cyberspace

The three-dimensional expanse of computer networks in which all audio and video electronic signals travel and users can, with proper addresses and codes, explore and download information.

Cyberstore

Shopping sites on the Internet.

Electronic mail (or e-mail)

An application that allows a user to send or receive text messages to or from any other user with an Internet address.

Encryption

Scrambling of data into an unreadable format.

Finger

A program that displays information about a particular user or users.

Firewall

Barriers that prevent outsiders from getting access to internal systems.

Flaming

Sending an abusive, harassing, or bigoted e-mail message.

Frame

A way to display multiple images on a screen.

FreeNet

Non-profit community organization that provides free access to e-mail and information services and to computer networks such as the Internet.

FTP
> File transfer protocol – allows transfer of files between two computers.

Geek
> A hacker seen in negative stereotypes (i.e., antisocial, obsessed with computer systems).

Gopher
> A system that lets you find information on the Internet. It can "talk" to other gophers; gophers can also link to other data by "tunnelling" through the Internet.

Hacker
> A person who enjoys exploring or even manipulating computer systems.

Home page
> The first page of a Web site.

Host
> A computer with full two-way access to other computers on the Internet. A host can use virtually any Internet tool, such as WAIS, Mosaic, and Netscape.

Hypertext
> A link between one document and other related documents elsewhere in a collection. By clicking on a word or phrase that has been highlighted on a computer screen, a user can skip directly to files related to that subject.

Hytelnet
> A menu-driven directory of publicly accessible Internet resources.

Internet
> An open global network on interconnected computer networks that enables computers of all kinds to share services and communicate directly.

Internet Service Provider (ISP)

An ISP offers dial-up access to the Internet, usually for a monthly fee.

IRC

Internet Relay Chat—a communications program that allows real-time conversations among multiple users; similar "chat" functions are the meat-and-potatoes of on-line services like America Online.

Java

A programming language based on C++ that will run on a wide range of computers.

Jughead

A tool used with Gopher. Name pays homage to Archie in the comic strip of same name. Sister tool, Veronica, searches all of Gopherspace; Jughead searches the files of just one gopher.

Lurking

Hanging around a mailing list or newsgroup without contributing to the discussion; sometimes used as a means of learning before becoming an active participant.

Modem

Contraction of mo(dulator) and dem(odulator); an accessory that allows computers and terminal equipment to communicate through telephone lines or cable.

Newsgroup

An electronic bulletin board where you can post messages, read messages, and add comments.

Netiquette

Internet etiquette.

Newbies

Users new to the Internet.

Node

Any device that is connected to a network. On the Internet: a "synapse" that stores or relays data that are moving down the line.

Phreak

Hacker who specializes in getting into telephone systems.

Ping

A program that checks to see whether you can communicate with another computer on the Internet.

Scrog

To damage, trash, or corrupt a data structure.

Shareware

Low-cost or free software that can be shared by Internet users.

Sites

Electronic locations where a person can log on and view an estimated 5.1 million screens of information.

Surfing

Exploring the Internet.

Telnet

A telnet command initiates a connection to a remote computer over the Internet. It allows you to log in to a distant computer and use it as if your terminal were directly connected to it.

Trolling

A controversial remark designed to draw flames.

Troughing

See Lurking.

Tunnelling

Enables information to be transmitted from one computer to another over a public network.

Universal access

The ability to get on-line to a network from anywhere or any place.

Upload, download

To move a file from one computer system to another.

URL

Universal Resource Locator—an address for a Web site such as http:/www.timeinc.com (which brings you to the opening screen, or home page, of Time Inc.'s own Web site, Pathfinder).

Usenet

A set of newsgroups considered to be of global interest and governed by a set of rules.

User

Someone doing "real work" on the Internet, using it as a means rather than an end.

UserID

A compression of User Identification, the unique account signature of an Internet user; that which precedes the @ (at) sign in an e-mail address.

Veronica

A tool used with Gopher (its name pays homage to the Archie comic strip) used to search for files located on computers that are Gopher servers.

Virtual reality

An interactive, simultaneous electronic representation of a real or imaginary world where through sight, sound, and even touch, the user is given the impression of becoming part of what is represented.

Virus

A cracker program that searches out other programs and "infects" them by embedding a copy of itself in them.

World Wide Web (WWW)

Internet service that gives universal access to a large universe (web) of documents.

Yahoo!

A search engine that can be used to find information on the World Wide Web. Other search utilities such as Excite provide the same service.

About the Authors

John (Jack) G. Nellist

As general transmission engineering manager at BC Telecom, Jack led the development and implementation of fiber optic technology in the British Columbia Telephone network. He was appointed coordinator for the first field trial of a fiber optic transmission system in 1978. Jack presented one of the first papers on an operating fiber optic system in 1979 at the International Fiber Optic Conference held in Chicago, Illinois.

As a communications consultant, he participated in the marketing, design, and installation of many fiber optic projects provided by Fluor Daniel Telecommunications Services Division of Irvine, California, from 1984 to 1986.

He is the author of the book *Understanding Telecommunications and Lightwave Systems,* 2nd ed., New York, NY: IEEE Press, 1996.

He is the co-author of the book *Commercial Building Telecommunications Wiring,* New York, NY: IEEE Press, 1996.

Elliott M. Gilbert

Elliott has been involved in the engineering and business aspects of telecommunications for over 30 years. He was the data transmission systems engineer for GTE Lenkurt, manager of product planning for the GTE Fiber Optics Communications Division, and director of lightwave marketing at Harris Corporation.

Most recently, Elliott has been the consultant for a wide variety of telecommunications projects, including the design and implementation of a large metropolitan fiber optic network, and the preparation of the curriculum for a junior college fiber optic telecommunications course.

He is the co-author of the book *Commercial Building Telecommunications Wiring,* New York, NY: IEEE Press, 1996.

Index

For further information on these and other Artech House titles, including previously considered out-of-print books now available through our In-Print-Forever® (IPF®) program, contact:

Artech House
685 Canton Street
Norwood, MA 02062
Phone: 781-769-9750
Fax: 781-769-6334
e-mail: artech@artechhouse.com

Artech House
46 Gillingham Street
London SW1V 1AH UK
Phone: +44 (0)20 7596-8750
Fax: +44 (0)20 7630-0166
e-mail: artech-uk@artechhouse.com

Find us on the World Wide Web at:
www.artechhouse.com